Help Your Child with Maths

Alan Graham is a lecturer in mathematics education at the Open University. He has three children, Luke (12), Ruth (11) and Carrie (3). His wife, Hilary, is a lecturer in social policy.

Alan T. Graham

Help Your Child with Maths

Fontana/Collins

First published in 1985 by Fontana Paperbacks,
8 Grafton Street, London W1X 3LA

Copyright © Alan T. Graham 1985

Set in Linotron Times
Made and printed in Great Britain by
William Collins Sons & Co. Ltd, Glasgow

Conditions of Sale
This book is sold subject to the condition
that it shall not, by way of trade or otherwise,
be lent, re-sold, hired out or otherwise circulated
without the publisher's prior consent in any form of
binding or cover other than that in which it is
published and without a similar condition
including this condition being imposed
on the subsequent purchaser

Contents

Acknowledgements 7
Introduction 9

Part I The Preschool Years

1 Sums for mums – and dads 13
2 Describing with maths 22
3 Where does this one go? Naming and sorting 25
4 Buckle my shoe – beginning to count 39
5 Mine's bigger! Looking for differences 53

Part II The Primary Years

6 What maths is all about 67
7 Whole numbers 73
8 Some games and activities with numbers 89
9 Fractions 95
10 Decimals 108
11 Percentages 120
12 Basic arithmetic 132
13 Knowing what sum to do 141
14 Measuring 150
15 More measuring 164
16 Shapes and solids 178
17 A sense of proportion 195
18 Patterns 207
19 Diagrams and graphs 220
20 The X factor – algebra 227
21 She who dares, wins! 239
22 With a little help from my calculator 250

Answers to practice exercises	269
References	277
Index	281

Acknowledgements

I would like to record my thanks to a number of people who helped me in writing this book. First of all, Dr Helen Roberts (Bradford and Ilkley Community College) with whom I carried out a research project entitled 'Sums for Mums'. This project was funded by the Equal Opportunities Commission and was a major inspiration for me in deciding to write this book. My thanks to John Mason for suggesting three of the activities in chapter 5.

Special thanks are also due to Ken Tyler for many helpful comments and suggestions on an earlier draft. They were much appreciated. I would like to thank the various people who contributed to the typing of the manuscript. The bulk of the first draft was typed by Debbie Skeats, while it was Debbie Bailey who put together the final draft.

Most of the cartoons were drawn by Anson Moss who always seemed to know what I wanted even when I couldn't describe it.

Introduction

This book has been written in two parts. The first deals with maths in the preschool age range. After reading this I hope you will be encouraged to find that much of what your toddler is learning and playing at is in some sense mathematical. Part II, the main section, deals with the sort of maths which your child will tackle in the primary school.

The book has been written with two central aims. First of all it provides you with a course in basic maths. This will give you a good grasp of the mathematical principles which your child will be grappling with in school. Secondly, and more importantly, it contains suggestions for the sort of activities and conversations at home which will encourage your child to be curious about mathematical things and to see maths as both challenging and fun. A major factor here is your own enthusiasm, so sharpen your pencil and be prepared to get stuck in!

Part I

The Preschool Years

1
Sums for mums – and dads

What memories of school still linger in the cobwebs of your mind? For many there may be surprisingly few because it seems that the really powerful memories relate to events outside the classroom – a first love, the death of someone close, a special song. But this is not always so. What are your memories of the mathematical experience? Odds on you're saying to yourself 'Hopeless! I could never do maths' or 'Terror! My mind used to go blank with fear.' If many people are mathephobic, that is, math-hating (and research evidence suggests that they are), then you might be asking questions like:

- Why do most children hate maths lessons?
- Could I succeed now at maths?
- Can I help my child enjoy maths?

If so, then read on . . .

Why did I hate maths at school?

There are many reasons why children learn to hate maths, but you may have noticed that women tend to have worse memories of maths lessons than men. Being good at maths, it seems, is all part of being rational, competent and manly. Being a woman and having female hormones, it is often argued, doesn't have very much to do with mathematical ability.

Well, as mathematicians say, this is just a load of spheres. On the contrary, I believe that the reason that so many women are mathephobic is not because of their nature but because of the way it was presented to them at home and at school. Moreover, this fear of maths that girls often experience is also shared by many boys. Recently I received several hundred letters from mothers, many of whom described their experiences of maths at school. What was remarkable was that they mostly concentrated on the sadistic tendencies of the maths teacher, rather than on the subject itself.

> If you moved a finger or fidgeted while she was talking she would pounce on you, gather up all your books, pencils, satchel, in fact everything belonging to you, and in a wild rage throw the lot out of the window.

> . . . involved a lot of shouting and tears to such an extent that fear of the scenes to come rendered me unable to comprehend the simplest explanation.

As this correspondent added: 'Maths was a subject to be dreaded – not because of maths, but because of Miss Nish.' Sadly, maths lessons can sometimes become arenas of stress where the teacher exerts power and control over the pupils. Knowledge becomes a stick to beat the children with rather than something about which to provoke their curiosity. Girls seem to survive these pressures just as well as boys up to the age of about twelve, but from then on their confidence and performance go downhill. Perhaps this is because they think it is unfeminine to be mathematical, or they don't think maths will help them to get a job. Or maybe they just freeze up under the stress of maths lessons.

Can I really help my child with maths?
Of course you can. Sometimes parents get put off helping their child in case they 'teach them the wrong method'. Well, maths isn't just about methods of doing sums, it is also about ideas, patterns, logic. . . . All these things can be encouraged in your child without treading on the teacher's toes. (Helping your child to ask the right question can be more valuable than concentrating on getting the right answer.) And if you are unsure about methods, don't be afraid to ask your child's teacher. They're not all dragons like Miss Nish.

Another worry which some parents may have is of exposing their children to certain maths topics out of their 'natural' sequence. You can be reassured on this question also. Although mathematics is sometimes regarded as a highly linear subject where each topic builds neatly on the one which precedes it, it is certainly the case that children do not learn it in this neat, ordered fashion. On the contrary they rarely, if ever, have a complete grasp of topic X before moving on to topic $X + 1$. So don't be amazed if your child shows a less than perfect understanding of something you thought was quite clear. She may have acquired her own understanding which meets her own needs (like knowing enough to convince her teacher that she understands it, for example!).

Will my maths be up to it?

The inspiration for writing this book was derived from a research project carried out in Bradford by myself and Dr Helen Roberts in 1982. We devised and piloted an experimental course in basic maths for mothers who wanted to be able to help their children. The course was called 'Sums for Mums'. The question of whether their maths would be up to it was also at the forefront of their minds. Clearly in order to be of help to your child you need to have a reasonable grasp of the concepts yourself.

The maths course presented in part II of this book is based on the work carried out by the Bradford mothers. At the end of the course all mothers felt that they had achieved a good understanding of the basics. Here are some of the facts which seemed to contribute to their success:

CONFIDENCE
> I felt I could ask the sort of question that I wouldn't have dared ask at school: like 'What is a decimal?'

The women felt able to take a mature approach and be honest in admitting they didn't understand.

RELEVANCE

> Yes, I think you can associate more with it. When you get to our age, when you've a family and a home, I think you can associate more if you do put it to more practical things. You can see it better in your mind's eye.

They were able to see how the various maths ideas related to the world of home and work.

MOTIVATION

> I thought I might be able to help Helen because it was always her dad that did.

> I was telling the girls I work with about this course and they turned round and said, 'What do you want to do maths for?' You know, as if it's something a woman shouldn't do and I thought, I felt really indignant about it. So this has spurred me on more. . . .

The mothers had chosen to take the course. They wanted to learn maths both for their own sakes and to be able to help their children.

UNDERSTANDING

The mothers were very different now to what they were like when they were fifteen. Now with a broader vocabulary and experience of life they were quickly able to grasp concepts they had never understood before.

ENJOYMENT

> I found all of it interesting. I'm wanting it to go on and on.

> I shall miss it when it's done.

Are you doing another one next year? We'll join again.

SENSE OF HUMOUR

Learning maths can be fun. So can teaching it, as I discovered from these lively women.

Hasn't maths changed a lot since I was at school?

Maths has changed a bit since you were at school but not by as much as you think. The changes have occurred more in the language of maths than in the topics covered. Basic arithmetic is still central to primary maths and today's twelve-year-olds are still having as many problems with decimals and fractions as you did. Chapter 6 takes a closer look at the question of 'What is maths?'

Will it help to use a calculator?

While it might alarm you at first, the answer to this is yes. About three quarters of adults have access to a calculator, but perhaps like you many may be frightened of them and would never use them. This book should help you overcome 'calculator anxiety' and indeed you will see how the calculator can be used to teach maths to yourself and your child. If you don't own a calculator, borrow one as you work through the book. Indeed you could buy your own calculator for about the cost of this book!

How will this book help me to help my child?

Whatever way you choose to help your child, there is one essential ingredient which must be present – *fun*! If it becomes a chore or a battle then stop what you are doing and try a different approach. If you watch young children at play – for example, cooking the dinner in a cardboard-box oven or driving an imaginary car – it is clear that they only have a vague understanding of what they are doing. The same is true when they are learning maths, so don't expect them ever to have a complete 'adult' grasp of a mathematical concept. As a parent your main contribution will be to share with your child the pleasure and challenge of the *process* of doing maths. If you try to focus on an *end product* (getting all the sums right) there is a good chance that you will transfer your own anxiety about maths to your child. It should be stressed therefore that it is not intended that this book should be used as a maths textbook, with you in the role of teacher and your child as pupil. This could be a recipe for mathephobic disaster for you both! The maths exercises are there for you to work through to provide you with a confident grasp of basic mathematical concepts. However, each chapter also contains a number of questions and practical activities which you can use to stimulate your child's thinking and curiosity about mathematical ideas. The ideal setting for these conversations might be in the kitchen, in a supermarket or on a long journey.

How, then, can parents encourage in their children a curiosity and excitement about mathematical ideas? Below are a few general pointers to indicate the sort of things you might say and do with your child to achieve this aim by creating what professionals call a 'mathematically stimulating environment'.

- Where possible try to respond to your child's questions positively.
- Encourage in her the desire to have a go at an answer, even if the answer is wrong. Wrong answers should be opportunities for learning, not occasions for punishment!
- Try to think about what makes your child tick and the sort of things she might be curious about.
- Resist providing easy adult answers to your child's questions but rather try to draw further questions and theories from her.
- If your child gives a wrong answer it is probably for a good reason. Try to discover the cause of the problem – it could be, for example, that she is simply not ready for the concept.
- There may be specific educational toys and apparatus which are helpful to have around. Having said that, a four-year-old will learn to count as effectively or ineffectively with pebbles or bottle tops as with 'proper' counting bricks!

- Finally, remember that 'how' and 'why' questions are at a higher level of curiosity than 'what' questions and should be encouraged. (For example, *'why* does $2 \times 3 = 3 \times 2$?' is a more stimulating question than 'what is 2×3?')

Can I help my child who is hopeless at maths?

Many parents who were unsuccessful at maths themselves have the depressing experience of watching their children follow in mother's or father's footsteps. Being hopeless at maths seems to run in families: or does it? While it certainly seems likely that *confidence* in maths passes from parent to child, it is less certain that 'mathematical ability' is inherited in this way. So if you consider yourself to be a duffer at maths, there is no reason to condemn your child to the same fate. Indeed, there is much that you can do to build your child's confidence and stimulate her interest in the subject.

The biggest barrier to learning mathematics is *fear* – the fear of being shown up that you don't understand something which seems to be patently obvious to the rest of humanity. The best way of helping your child is to start by trying to overcome that fear in yourself. It doesn't matter if you don't know a simple fraction from a compound fracture; your competence at maths is less important than your willingness to be honest about what you don't understand yourself. However bad you think you or your child are at maths, this book will help. Firstly, it will give you a better grasp of basic maths yourself – essential if you want to provide information for your child. The many explanations of mathematical ideas and structured exercises which are contained in the book will give you a bedrock of understanding and confidence so that you feel equipped to help your child. Secondly, the book offers insights about the sorts of mistakes and confusions which children often experience. You will discover that your child isn't simply wilful or stupid but may be making errors which are consistent and 'reasonable'. Finally, you are offered a range of simple everyday activities – games and ideas for conversations – which will complement and reinforce the maths learnt in school and which should stimulate curiosity and excitement about mathematical ideas. For example, there are several entertaining puzzles in chapter 8 and some calculator games in chapter 22.

To what age of child does this book relate?

The maths in this book contains the main topics that a child should cover up to the age of eleven or twelve years. The main section of the book is called 'The Primary Years' and covers the five- to eleven-year range (chapters 6–22). This is preceded by chapters 2–5 on 'The Preschool Years'.

Should I do some maths myself?

Well, just as with cooking, carpentry or playing the piano, learning maths doesn't happen just by reading a book about it. Although explanations can lead to understanding, practice is necessary if you are to achieve mastery. If you aim to be able to help your child with maths there are therefore two good reasons why you should tackle the exercises in this book.

1. Doing maths is the best way to understand it yourself.
2. If your child sees you doing and enjoying maths problems, she is more likely to do so herself.

These exercises begin in chapter 3 and continue throughout the rest of the book. They come in two types. Those which are labelled A, B, C . . . have been designed to let you 'discover' something about maths. The sort of questions asked here will be:

- What do you notice?
- Can you see a pattern?
- Does this suggest a general rule?

and so on.

Then at the end of most chapters you will find a set of practice exercises, labelled 1, 2, 3 . . . These will help you to consolidate what you have just learnt. Both the 'discovery' and the 'practice' exercises are equally important, so please resist the temptation to skip over them!

Not surprisingly the level of maths in the preschool section is fairly basic. However, there are a number of exercises here also and you are warmly encouraged to do them.

These exercises will help in several respects by:

- consolidating your understanding of what you have read
- relating the theory to the practice of helping your own child
- providing you with ideas for specific activities which you can do at home to help your child think mathematically

2
Describing with maths

If you ask several professional mathematicians to summarize the basic ideas of maths they almost certainly won't come up with the same perspective. The mathematical ideas which run through part I of this book are offered with young children very much in mind, even though the ideas are also of central importance to maths at all levels. They are all linked to the very simple idea of *describing* and focus on how children learn to make sense of their world by observing similarities and differences.

Mathematics, therefore, isn't just about crunching numbers. Your child will be tackling some of the most important ideas in mathematics long before she knows that two twos are four.

Ways of describing

Describing the objects, the people and the relationships around us is an important human activity. As we mature from early childhood our descriptions become more subtle and complex. When Carrie was eighteen months she was at the *naming* stage. (Incidentally, throughout part I the name of Carrie will keep recurring. She happens to be the two-year-old in my life at the time of writing.) She was fully occupied learning the names of her brother Luke and her sister Ruth, and a long list of everyday objects – ball, teddy, dog, window, and so on. Not surprisingly her conversation was a bit limited as her descriptions were restricted to identifying the name of the object in question.

By the age of two Carrie has moved on to *sorting* objects and people into useful 'sets'. She is discovering gender, for example ('Carrie is a girl', 'Luke is a boy'). She is struggling with colour ('The car is red', 'The cup is red') and she is confident in matching objects to members of the family ('This is Luke's', 'Ruth's got Alan's shaver').

Even at this stage Carrie is discovering that number is an interesting way of describing things ('Look, two tractors!'). Although she isn't yet engaged in formal *counting* she is beginning to recognize pairs and threes and to use the words two and three to describe situations.

When we count we are usually dealing with things which are similar in some way. However, as well as describing similar things, Carrie is also noticing differences. Soon she will be talking about 'bigger' and 'smaller' and choosing the 'biggest' portion of chocolate. The mathematical process she is engaged in here is *comparing*. Although many types of comparisons are available, it is the comparisons of size and shape which tend to be particularly mathematical.

Comparing usually involves looking at just two objects. It will be some time before Carrie can look at five or six objects of different size and put them in *order*. Nicky, who is five, is already able to order as many as ten objects by size (fitting different-sized beakers inside each other, for example).

Nicky is also coming to the most sophisticated process listed here – *measuring*. It will be some time before he is taught about the standard units of measure (grams, pence, centimetres, and so on) but he is hearing them every day and beginning to learn about what they mean. Even Carrie has made a start in this direction with her tape measure. All visitors to her house must remove their shoes and have their feet measured. Interestingly, they all measure the same – 20p. My finger, recently measured, came to '10p and a half'. One reason that measuring is more difficult than counting is that it often produces answers which are not whole numbers. For example, when Nicky stands on the scales, his weight will not be exactly 22 kg but, perhaps, twenty-two and a bit (later twenty-two and a half) kg. Although many children, like Carrie, enjoy playing with measurement at the preschool age, formal measuring skills are difficult and are left to part II of this book.

In this chapter I have listed six ways of describing things which most preschool children use at some level and which are key ideas in mathematics.

Here they are again:

naming; sorting; counting; comparing; ordering; measuring.

The diagram below shows how they relate to each other.

Ways of describing

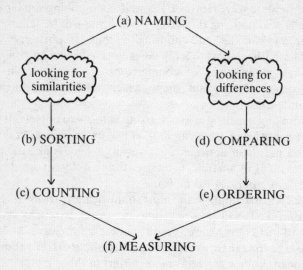

The next three chapters look more closely at these forms of mathematical descriptions and particularly at how they can be encouraged in your preschool child.

3
Where does this one go? Naming and sorting

Bina, who is eight years old, hadn't been making good progress at school. Eventually her teacher called in the educational psychologist to have a session with Bina and make an assessment. She was shown a model farmyard and given a collection of various farmyard animals – sheep, cows, goats, pigs; and so on. The educational psychologist then asked Bina to sort the animals into the fields and he settled down to observe her and make notes. Poor Bina – what a failure! She put three cows, one goat and one dog in the first field, two pigs and four sheep in the next. The psychologist was writing up his notes on Bina's inability to sort simple shapes when, tired of playing with the farmyard, she asked 'Can I put them away now?'

'Oh yes, do,' replied the psychologist and continued with his notes. A few minutes later he looked up to find that Bina had put all the pigs into one box, the cows into another and so on. He stopped writing and asked, 'Tell me, Bina, why did you put different animals all together in the same fields?'

'Well,' replied Bina, 'I thought they would just like to talk to each other.'

We tend to think that children have a poorly developed understanding of the adult world. What we forget is that our own understanding of *their* intentions and expectations is just as faulty, if not more so. It is easy enough to observe what children do and say, but why they do it and what it means to them is slippery stuff indeed. This chapter starts at the early stages of a child's language development with the most basic form of description – that of *naming*. It goes on to look at three further skills necessary for a child to make sense of patterns and similarities in the world around her. These are *matching, sorting* and *classifying*.

Naming

The naming of people, objects, experiences is where a child's language first takes shape. Until she can use the agreed labels or names which we attach to the various bits of our world, it is difficult for the child to show much evidence of mathematical understanding. By eighteen months to two years your child will probably have mastered an impressive vocabulary of nouns:

> dog, toes, cow, cat, finger, tummy . . .

What takes longer is the realization that some of these words are names of animals, while the remainder are parts of the body. The first step is to get her thinking about the properties or characteristics which make a ball a ball and not a dog or a brick. Later she may be able to work backwards from a description of the properties to identify the object. Here is an example of how this sort of thinking could be pursued.

PARENT: [*Holding up a toy elephant*] What's this?
ISABELLA (aged three): An elephant.
PARENT: How do you know it is an elephant?
ISABELLA: 'Cos it's got a trunk.
PARENT: Yes, it has – a big trunk. What else has the elephant got?
ISABELLA: . . . got a little tail.
 . . . and a big fat tummy.
PARENT: Yes – and what about its ears?
ISABELLA: Big ears.
PARENT: Yes, big floppy ears. And what colour is the elephant?
ISABELLA: [*Hesitates*] Black.
PARENT: Yes, nearly black. It's grey really. And do you know what those things are there? [*Points to tusks*]
ISABELLA: [*Shakes head*] Teeth?
PARENT: They are a bit like teeth. Those are called tusks. The elephant uses his tusks to poke things with.
Now do you know what this is? [*Holds up a toy tractor*] etc.

Isabella knows more than just the elephant's name. She has noticed some of the characteristics which make it an elephant and not something else. This conversation has drawn her attention to one or two more features which are characteristic of an elephant. She will discover that there are other animals which are grey, some with large floppy ears and still others with tusks. But not many animals have all of these characteristics. And of course the trunk is the clincher!

The next stage is to play the game backwards. One version of this is called 'In the Bag'.

IN THE BAG
Hide several objects in a bag, put your hand in and say 'Can you guess what I have here? It's an animal. It is grey. It has four legs. It has big ears. . . .'

The child can guess at any stage and when she gets it right the object is revealed and the next round begins.

Clearly it is more fun if you start with the least helpful clues (i.e., leave the trunk to the end!). It won't be long before your child will want to do the describing herself and have you guessing the hidden object.

Matching

There are a number of words whose meaning in mathematics is rather different from their everyday use. Matching is such a word. 'Everyday' matching suggests putting together things which are similar in some way (matching colours for example). In mathematics, matching means getting the correct number of saucers to cups, dolls to prams, riders to horses. So, in maths we are concerned with matching the *number of objects* (e.g., matching pairs).

Matching in mathematics

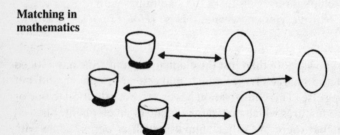

The more mathematical meaning of matching (matching numbers) will be dealt with in the next chapter under the heading 'one-to-one correspondence'. In this chapter I will use the word matching in its everyday meaning – putting together two things that are similar.

Matching, then, is a simple form of sorting – putting together identical objects. Here are some examples where children are encouraged to try their hand at matching.

Example 1

TABITHA: [*Hands adult a knife*] A knife.
ADULT: Oh thank you, Tabitha. Can you find me another knife like this one?

[*Tabitha makes a pile of knives*]

Example 2

[*Carrie is playing with a bag of plastic clothes pegs*]

ADULT: [*Holding up a yellow peg*] What colour is this peg, Carrie?
CARRIE: Red. [*All colours are red to Carrie*]
ADULT: No, this one's red [*points to red peg*]. My peg is yellow. Can you give me some more yellow pegs like this one?

[*Carrie then proceeds to match the yellow pegs successfully and hands over four or five of them*]

There are a number of enjoyable matching games on the market of which here are a few examples.

- Posting toys, where the child learns to find the hole which matches the appropriate shape before 'posting' the shape into the box. Next time your child uses a posting toy, try to help her learn to name the shapes of the holes. If you are unsure of them yourself a list is included on page 34. Incidentally, these posting toys can be homemade using an empty ice-cream carton for a box and the holes cut from the lid. It is more challenging to find suitable shapes to fit through the holes but spools or corks could be used for circular holes and dice or small wooden bricks for square ones.
- Picture dominoes – these usually have number dots on the reverse side so the game can later be upgraded to a form of real dominoes.
- Picture lotto, where each child is given a board with nine familiar pictures. A sequence of picture cards are drawn which will match only one child's picture. A child then places the matching card on her identical picture until the board is full. Bingo! As well as developing the notion of matching, this sort of game is an excellent way of extending

the children's vocabulary and this takes us back again to the previous stage of naming.

Matching activities needn't just involve pictures. Children from two or three years can try their hand at matching shapes, colours and sizes. Then later they can play matching games with letters (both capital and lower case, i.e., 'small'), number dots and numerals. There are plenty of educational games on the market which practise these sort of skills. But really, most of these activities can be set up in the kitchen with 'games' like choosing the right saucepan lid for the saucepan, finding matching pairs in a pile of white socks or tidying up the jungle of family shoes under the stairs. (It is one of the ironies of parenthood that when children are young enough to be willing to play tidying games they usually end up making a bigger mess than they started with. By the time they are old enough to be of real help the tidying game has strangely lost its appeal!)

To end this section on matching here is an activity to get you thinking about the skills demanded when a child uses a posting toy.

Exercise A

Carrie at eighteen months was playing with her new posting toy. It was a fairly simple model with only three shapes and corresponding holes, as follows:

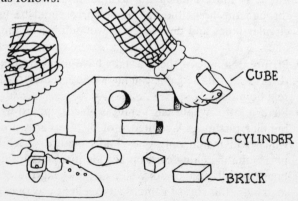

WHERE DOES THIS ONE GO? NAMING AND SORTING

(i) Which shape do you think Carrie learnt to post first?

cube

(ii) Why? *because she's holding it*

COMMENTS ON THIS EXERCISE

Like many parents, I experienced a sense of wonder as I watched Carrie struggle with her new posting toy. Her concentration and determination to solve the problem were remarkable. At first she tended to pick up a shape in a random way and try to shove it into any and every hole. However, within just a few days she was ignoring the cubes and bricks and was picking up only the cylinders. These shapes were being posted through the circular hole with great skill and pleasure. It seemed to me that the reason Carrie found this shape easiest to post was related to the skills needed to post a shape successfully. They are as follows:

- being able to *recognize the shape* and match it with the corresponding hole
- knowing to *turn the shape* so that it is in the correct orientation to go through the hole

Looking at the second of these skills I realized why the cylinder proved to be the easiest of the three shapes – with a circular hole the orientation doesn't matter. As I expected, Carrie mastered the cube next, since there were six possible faces and four different orientations which would work.

She is now posting much more complex and varied shapes but I would suggest that your child's first posting toy should be very simple indeed.

Sorting

We have already stressed the importance of looking at objects in terms of their properties and characteristics. Clearly there are as many ways of sorting as there are characteristics. Before looking at some of these, pick up a pencil and jot down as many as you can think of in about ten seconds in the space below. You may find it easier on a particular object – like a banana for example.

Exercise B

Jot down here the characteristics of a banana

[handwritten: yellow, like a boat, slippery]

COMMENTS ON THIS EXERCISE
Here are some of the words which I wrote down:

yellow, smallish, crescent-shaped, soft, sweet smell

These words include particular examples of the most common attributes which we use to describe things. These are:

- colour
- shape

and - size

There are, of course, many other ways of describing things, such as smell, texture, taste and so on. However, the two which are of greatest importance in mathematics are shape and size. Two other features which are useful descriptions in maths are position (*where the object is*) and orientation (*which way round* it is). It is worth noting that whereas shape and size are permanent properties of a particular object, position and orientation are not. Position and orientation can be changed without causing a change in the object itself. Thus a brick can be moved and turned round and it is still the

WHERE DOES THIS ONE GO? NAMING AND SORTING

same brick. Although this fact is obvious to adults it is less so to children – a point that will be taken up again in the next chapter under the heading 'Conservation of Number'.

Exercise C asks you to think about a wide range of the attributes we use when describing and to make a list of the sort of words which would help children enrich their descriptive powers. You may find it helpful to keep a particular object in mind – a car, for example. One or two examples have been supplied to get you started.

Exercise C

Attribute	Examples	Jot down your own examples
Colour	red, green	
Shape		
Size		
Smell		
Texture	rough	
Sound		
Position		
Orientation		

SOME COMMENTS ARE GIVEN ON PAGE 37.

Here are a few other useful words which are likely to crop up when your child is talking about sorting:

different, like, alike, same, another.

Clearly the same set of objects can be sorted in many different ways (the red ones, the round ones, the big ones, and so on). Let us for now concentrate on the mathematically important idea of shape. Here are some of the most common shapes and their names. It is a good idea to get your child used to these names as soon as she can recognize the shapes. It will give you both something to talk about during the long hours of posting!

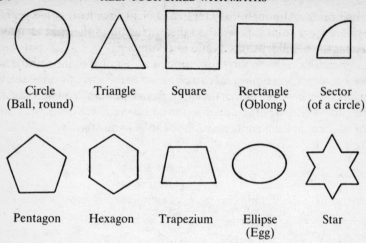

Written in brackets below some of these mathematical terms are a few of the words which your child may prefer to use – for example, ball rather than circle. This raises a difficulty which you may have already come across. The problem is that the shapes as represented by the holes in the box are two-dimensional, whereas the shapes themselves are solid and in three dimensions. For example, the hole which your child may call a ball is really a circle in two dimensions. Although a ball of suitable size could be pushed through this hole, the relevant shape is probably a cylinder like this:

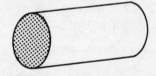

Your child won't be too worried by this and neither should you, since the concepts involved are well beyond the average two-year-old. She will pick up your words and at first you can use some of her words to describe the shapes. Build up gradually and don't worry if at first she gets some of them wrong; there is plenty of scope for refinement later on.

WHERE DOES THIS ONE GO? NAMING AND SORTING

One way of getting older children (say around four to five years old) to start thinking about the properties of shapes is to play 'Feelies'. A child is asked to feel a shape which she cannot see (perhaps blindfold the child or put the object in a bag or behind her back into her hands). She feels the object, describes what she feels and then says what it is. For example, she might say, 'It's got corners . . . and flat sides. . . . It's a cube, I think.' Another way of developing sorting skills is to play 'Odd One Out'. For example, from:

- four red bricks and one yellow one
- four round shapes and one triangle
- four things to eat and one brick
- four garments and one spoon

and so on.

Exercise D asks you to do some sorting yourself. When you have completed the table below you may find it useful when planning sorting games with your child like 'Odd One Out'.

Exercise D

Write down as many household objects as you can which have the following shapes or properties.

Shape or property		Item
shapes	Circle	
	Square	
	Rectangle	
	Red	
	Shiny	
	Makes a noise	
	Silent	

COMMENTS ON THIS EXERCISE ARE GIVEN ON PAGE 38.

Ian who is four is already enjoying sorting games which involve more than one property. For example:

> What is round and makes a noise?
> What is red and shiny?

and so on.

Although the four-year-old may be able to handle these questions in his head, the three-year-old may prefer to use her hands and sort things into piles. A loop of string on the table can be used to represent each 'set' of objects and the child then places the objects in turn into the appropriate loop. There are several useful ways of representing this sort of activity on paper in a diagram. However, these are not helpful for the preschool child and will be included later in chapter 19.

Classifying

When she is able to name the set or class of objects inside her loop of string your child is showing her ability to classify. However, sometimes her vocabulary may let her down. For example, look at this list of household objects:

> wooden spoon, egg beater, mixing bowl, measuring cup

They are clearly not all the same but your four-year-old may see that they share a common property and could be put into the same set. Classifying involves finding a name for the set. The adult label might be 'cooking utensils' although the child might say, 'They are all things you use in cooking.'

Children love to match, sort and classify and there is plenty of scope for these activities when tidying up round the house. Here are a few examples:

- laying the table
- putting away the cutlery
- pairing up the socks
- putting out the milk bottles

WHERE DOES THIS ONE GO? NAMING AND SORTING

- tidying up lots of shoes
- putting away laundry into correct drawers
- tidying away the toys
- putting away the groceries

Summary

The diagram below summarizes the main mathematical ideas contained in this chapter.

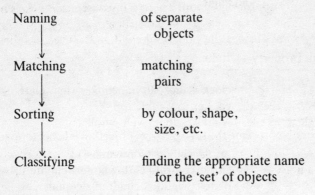

ANSWERS TO EXERCISES FOR CHAPTER 3

Exercise A

No comment.

Exercise B

No comment.

Exercise C

Attribute	Examples
Colour	red, green, yellow . . .
Shape	round, square, triangular . . .

Exercise C

Attribute	Examples
Size	large, wide, narrow, tall . . .
Smell	acrid, musty, rotten eggs . . .
Texture	rough, smooth, hairy . . .
Sound	loud, soft, shrill . . .
Position	high, low, behind, above . . .
Orientation	right way up, upside down, rotated through an angle . . .

Exercise D

Shape or property	Item
Circle	clock, saucepan lid, cooker ring, plate . . .
Square	cooker top, table top
Rectangle	window, door, wall, book . . .
Red	all *my* crockery! . . .
Shiny	my saucepans (sometimes!) . . .
Makes a noise	clock, central heating boiler, children . . .
Silent	plates, window, children sleeping (nearly) . . .

4
Buckle my shoe – beginning to count

'I count!' says Timmy. 'One, two, three, four, five, six.'

Well done, Timmy – not bad for two years three months. But is he really able to count? What does being able to count actually mean? If asked to count five sweets in front of him Timmy is as likely to say three as six. So while he has acquired some counting skills he still has some way to go before he can count accurately and really understand what it means.

This chapter looks at the skills which a child needs to have grasped in order to be able to count accurately. Later in the chapter we will look at the more complex aspects such as representing numbers and counting with larger numbers. But let's start at the beginning.

From sorting to counting

The previous chapter focused on the skill of sorting. It was stressed that we sort because we are interested in finding similarities and patterns. So once the red Smarties have been sorted into one pile and the green ones into another, what then? What else is there to say about these piles of Smarties? Well, perhaps that there are *lots* of red ones and *a few* green ones, only *one* pink one and *no* black ones. There are also *more* red ones than green ones in my box. If there is any eating to be done, most children develop an interest in these questions of 'how much' and 'how many' very quickly. For the first question in this chapter try to think about these two expressions 'how much' and 'how many' and how we use them.

Exercise A

ANTHONY (aged seven): How much sweets did you get?

Why should it be 'how many' in this sentence and what is the difference between 'how much' and 'how many'?

COMMENTS ON THE EXERCISE
We usually talk about 'how many' when there is a set of distinct items which can be counted (cars, people, trees, calculators, and so on). However, if we are dealing with quantities which can't be counted (amount of water, sand, pudding . . .) but which must be apportioned, we say 'how much'. Of course, we only have to cut our bread into convenient units such as slices or sandwiches and then we can use 'how many'. Not surprisingly children often confuse these two expressions.

Counting is a way of measuring *how many* items there are and therefore counting occurs when there is a collection of separate objects. For many years educational psychologists have been interested in how children learn to count and why young children fail to count accurately. Although I personally take many psychological theories about what children can and can't do with a large pinch of salt, one important concept, known as the conservation of number, does seem to separate the way we think from that of young children.

Conservation of number

As adults it is patently obvious to us that ten sweets are ten sweets no matter how they are arranged on the table. Similarly, a pint of milk does not change in quantity when it is poured from a bottle into a saucepan. In other words, we are happy to accept that the quantity of something is conserved (i.e., retained) even when it is moved in position or changed in shape. This belief, known as the principle of conservation, is not shared by young children. They see no reason to assume that the number of sweets is still ten, whether they are put on to a plate or scattered across the table. A possible explanation for this may be that their notions of 'how many' and 'how much' may be not quite the same as ours. For example, they may say that three large buns is 'more' than four small buns. And of course in the sense of how much there is to eat, they are right.

With very young children there is also evidence that when an object is moved into a different position it becomes, to them, a different object. In other words, children tend to accord more importance to position than to shape. For adults it is the other way round. To us it seems obvious that, whatever position an object is in, its appearance remains the same. Or is it . . . ?

Exercise B

(i) Write down the names of these three familiar shapes.

6   d   □

Name   _____   _____   _____

(ii) Now each shape has been rotated, as follows:

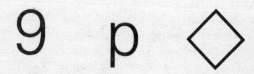

Is it obvious that these are still the same shapes?

THERE ARE NO COMMENTS ON THIS EXERCISE.

In the United States there are a number of preschool schemes designed to try to teach children to conserve. One has to wonder whether it is worth the effort since most children will come to accept this view of the world on their own by the age of six anyway without formal teaching. What is more important is that they learn to recognize what some teachers call 'the threeness of three'. In other words to realize that these sets shown below share the same property – their threeness.

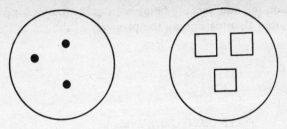

Indeed, recognizing the 'twoness' or 'threeness' of sets of objects will almost certainly occur before your child has mastered any of the formal skills of counting.

Before looking at counting skills, however, it is useful to make clear the distinction between what are known as 'ordinal' and 'cardinal' numbers.

Ordinals and cardinals

Most adults don't need to count their fingers very often to know that they have five on each hand (the number doesn't usually vary much over time). However, if you do count the fingers on one hand saying the numbers 1, 2, 3, 4, 5, you are aware that the number five represents a property of all the fingers taken together. You also know that neither the order in which you count your fingers nor the way in which they are arranged will affect your answer. It is still a 'bunch of fives' whatever way you look at it. For your child, finger-counting is less straightforward. For her the last object counted (in this case the thumb) is likely to be what takes on the property of 'fiveness'. Here is an example of a five-year-old who finds this confusing.

SIMON: [*Counting three beakers*] One, two, three.
TEACHER: [*Taking one away*] How many now?
SIMON: Three.
TEACHER: But I've taken one away.
SIMON: Yes, and that leaves two and three.

For Simon the numbers are being used as the names for particular items rather than describing the property of the set of items as a whole.

So five means to Simon the fifth finger counted rather than the total number of fingers. It's not surprising that Simon has found this confusing since we use numbers in a variety of different ways. Simon was concerned with the *order* of the numbers (first beaker, second beaker, third beaker, and so on).

When numbers are used in this way they are called *ordinals*. For example, the telephone number 812615 or the date 25/9/86 are making use of ordinal numbers. Numbers used to describe the *numbers of items in a set* (one, two, three, etc.) are called *cardinals*. Exercise C will give you some practice at sorting out ordinals and cardinals.

Exercise C

Tick whether ordinal or cardinal.

		Ordinal	Cardinal
(i)	The number on a footballer's shirt		✓
(ii)	The number of eggs in an egg box	✓	
(iii)	The attendance at a tennis match		✓
(iv)	The place a runner came in a race	✓	
(v)	Page number in a book		✓

COMMENTS ON THIS ACTIVITY ARE GIVEN ON PAGE 52.

What are the skills of counting?

Sam, aged two, is trying to count my fingers. 'One, two, ten,' he says. Carrie does a little better. 'One, two, three, four, five, six, seven,' she says, while at the same time vaguely points to my fingers.

Alice, who is four, comes very close. She manages to touch each of my fingers at the same time as she says the numbers. Unfortunately she has missed out my thumb, and started counting with my index finger as one.

All of these difficulties show just how complex the skill of counting really is. They also give some insight into the subskills which children need to have mastered before they are able to count accurately. The three relevant subskills are as follows.

Being able to:
(i) say the counting numbers (1, 2, 3, 4 . . .) in the correct order
(ii) relate the counting numbers to the objects being counted on a one-to-one basis. This means being able to say the counting numbers (1, 2, 3 . . .) at the same time as touching the objects being counted. This is known as one-to-one correspondence. (One-to-one correspondence has actually a wider meaning than this. It usually refers to matching up one set of objects [cups] with another set of objects [saucers].)
(iii) identify each of the objects to be counted once and only once. (This can be quite difficult with a large number of objects which are scattered in a random pattern.)

The next section looks at these three counting skills more closely and provides suggestions for developing them in your child.

Developing the skills of counting

(i) *The counting numbers:* the first and easiest of the counting skills is learning to say the number names in order. There is little that needs to be explained to your child here. These number words happen to have been chosen by English-speaking people and the choice of sounds is quite arbitrary up to the number twelve. From thirteen onwards the number words become a bit more interesting, but more of that later. Most children don't need to be taught the numbers from one to ten. They simply pick them up from the many counting songs and rhymes they hear. How many do you know of

BUCKLE MY SHOE – BEGINNING TO COUNT

these? See if you can jot down five numbers songs in the space below.

Exercise D

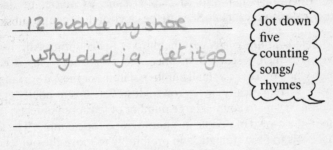

1 2 buckle my shoe
why did ja let it go

Jot down five counting songs/rhymes

COMMENTS ARE GIVEN ON PAGE 52.

(ii) *One-to-one correspondence:* when a child is just beginning to count at the age of two or three there is no need to go much beyond the number ten. What will need lots of practice, however, is the second of the counting skills listed in the previous section, known as one-to-one correspondence. What this means in terms of counting is being able to match up the saying of the counting words 1, 2, 3 and so on with the touching of the objects being counted. However, it is a more general skill which has everyday importance. Without one-to-one correspondence a child couldn't match up the correct number of cups to saucers, chairs to teddies or socks to feet. Much of children's play involves matching of this sort and it only takes a little adult intervention to get them talking about what they are doing. Here are some extracts taken from a tea party with a group of three-year-olds. The words which are in italics are part of our counting vocabulary and well worth stressing with your child.

- Oh dear, we haven't *enough* saucers.
- There are *more* cups than saucers.
- There are *less/fewer* saucers than cups.
- I think we've got *too many* cups.
- Now we have got *the same* cups and saucers.

Counting steps as you climb them is a good way of improving your child's grasp of the sort of one-to-one correspondence mentioned under counting skill (ii). Gradually your child will learn to bring into harmony the rhythm of counting and the rhythm of her feet on the steps. Encourage her also to clap in time to music, to dance rhythmically, to say 'tick tock' with the clock and, of course, to sing.

(iii) *Single-counting each item:* the third of the counting skills listed above is to include each item once and only once. This is easy enough when there are a small number of items or they are arranged for you systematically in a row. However, even adults have difficulties trying to count a large number of scattered objects – the number of a crowd of people in a photograph for example. The key to success is to be systematic and preferably to move the objects as they are counted. Unfortunately, children are not systematic. Furthermore, it is by no means obvious to them that even if you move the objects around they are still the same in number. This realization that objects have a permanence, irrespective of their position, only comes to children around the age of five to seven years. If your child is older than this and still seems to be having difficulties with this aspect of counting, here are a few suggestions which might encourage her to be a bit more systematic so that each item is counted once and only once.

- Let's put them in a row so that we don't miss any out.
- Next time start with the little finger.
- Put a mark on them as you count each one.
- Count them into my hands.

Representing numbers

'Learning your numbers' is considered a major milestone in a child's education. In the primary school this clearly means more than being able to recite the counting numbers in order. At some stage your child must learn to recognize and draw the symbols which we use to represent numbers:

1, 2, 3, 4, and so on.

These are called the *numerals*. My advice, however, would be to resist the urge to teach this important skill of number recognition at the preschool stage. There is so much about number that needs to be explored physically (handling bricks, cubes, conkers, pebbles, bottle tops, and so on) and orally (talking about bricks, cubes, conkers . . .) before complicated symbols like 1, 2, 3 and their friends are introduced. These key stages of development can be summarized as follows:

In the preschool and even infant-school years, the bulk of your child's learning should be at the 'do' and the 'say' stages. Resist the urge to rush your child into writing and recognizing symbols too early. Until your child has grasped thoroughly the 'threeness of three' principle by handling a range of different objects, the symbol '3' will have little meaning. However, one form of representation is appropriate and can be used even with very young children. This is to prepare, say, ten cards with, respectively, one up to ten dots shown on them. Try showing them to your child, building the number of cards up slowly, and simply say the number of dots on the card. It is a game which may last anything from twenty to thirty seconds at a time – when her concentration slips stop the game and try again later. Don't arrange the dots in any special pattern as your child may feel that this pattern has a special significance.

You might like to make one or two parallel sets of cards showing not dots but teddies or cars.

If this idea of introducing your toddler to a form of number representation appeals to you, you might like to have a look at the book *Teach Your Baby Maths* by Glenn Doman (see reference 1). This book explains how you can teach your child quite advanced number skills using the sort of cards described above.

Counting is fun!

Nicky, who is nearly five, finds counting easy and fun. He will count anything and everything and doesn't always use the 1, 2, 3 method. He is beginning to see patterns in number names – pairs, fives and tens particularly – and enjoys using these in his counting. One day he was exploring numbers above twelve. . . .

NICKY: . . . thirteen, fourteen, fifteen, sixteen, seventeen, eighteen, nineteen . . .
What comes after nineteen?
ADULT: Twenty.
NICKY: twenty-one, twenty-two, twenty-three, twenty-four, twenty-five, twenty-six, twenty-seven, twenty-eight, twenty-nine . . .
What comes after twenty-nine?
ADULT: Thirty.

Then Nicky's big sister got involved.

ANNE (aged six): Thirty, forty, fifty, sixty, seventy, eighty, ninety, tenty.

'Tenty' was a reasonable enough guess and Anne will learn soon enough about the exceptions. But both children are developing a strong sense of the pattern of numbers and using them as stepping stones to hop around. Later they'll be counting backwards in pairs and then wondering what happens after they get to zero . . . but more of this in chapter 7.

For most children, counting with large numbers is quite difficult. We tend to assume that the counting skills which seem secure for small numbers (one-to-one correspondence and conservation of number, for example) would be easily transformed to counting with larger numbers like 20, 50 and 100. However, this is not so. To some extent these skills need to be relearnt when the child starts to handle bigger numbers. If you can find a friendly four- or five-year-old, you might like to check this for yourself.

Exercise E

Find a willing guinea pig (human) aged around four to five years who seems able to count. Give her or him some counting tasks to perform, gradually making the task more difficult by varying the position of the objects and increasing the number of them. Bearing in mind the three counting skills described above and the idea of conservation of number, see if you can spot when the child's counting breaks down and why.

THERE ARE NO COMMENTS ON THIS ACTIVITY.

ANSWERS TO EXERCISES FOR CHAPTER 4

Exercise A
No comment.

Exercise B
No comment.

Exercise C

	(i)	(ii)	(iii)	(iv)	(v)
Ordinal	√			√	√
Cardinal		√	√		

Exercise D

How many of these counting rhymes do you know?
10 green bottles; 1, 2, 3, 4, 5, once I caught a fish alive;
10 little Indians; 5 currant buns in a baker's shop;
There were 5 in a bed; 5 little fish went swimming one day;
5 little speckled frogs sat on a speckled log;
1, 2, buckle my shoe; The 12 days of Christmas.

Exercise E

No comment.

5
Mine's bigger! Looking for differences

Like the last two chapters, the maths skills explored here are all linked to the central idea of describing. The skills which have already been covered – matching, sorting, classifying and counting – have all been concerned with describing objects in terms of their similarities. This final chapter in the preschool section looks at three important skills which children need in order to be able to describe clearly how objects *differ*. The skills covered are comparing, ordering and measuring. Bearing in mind that these are all ways of enabling children to describe more effectively, particular attention will be given to encouraging appropriate use of language.

Comparing

Just as matching involves putting together two objects which are alike, comparing means discovering what makes them different. For example, a child might match together two cubes from a pile of bricks because they are both cubes. But if she wishes to compare these two cubes she may observe that they are, perhaps, not the same colour, or that one is bigger than the other.

Exercise A

Below is an edited extract from a conversation between an adult and a four-year-old called Joe. They are playing with some buttons. As you read the extract, draw a line under any comparing word that is used. The first one has been done for you to get you started.

JOE: You've got <u>*more*</u> buttons than me.
ADULT: Yes, I've got fewer than you.
ADULT: Close your eyes. [*Adult removes a button from Joe's pile*]
  Now, have you more buttons or less buttons than before?
ADULT: Now you do it to me and see if I can guess.
ADULT: [*Holds up two buttons*]
  Which one is bigger?
  Is this one fatter or thinner?
  Is it wider?
  Are they the same colour? Which one is darker?
  Do they feel the same? Which one is bumpy? Yes, that one is smooth.

COMMENTS ON THIS EXERCISE
The comparisons being used in this button game were as follows:

  number (more/fewer)
  size (bigger/fatter, thinner/thicker)
  colour (darker/lighter)
and texture (bumpy/smooth)

Naturally parents are keen to develop their child's ability to describe and make comparisons with words. The comparisons that mathematicians are particularly interested in are those of number, size, shape and position. Some of the words used to describe shape have already been mentioned in chapter 3. Here are the common words associated with each of the other three concepts.

NUMBER
- many/more
- few/fewer/less
- some/a lot
- none

SIZE
- big/bigger; small/smaller
- large/larger
- long/longer; short/shorter
- wide/wider; thin/thinner
- narrow/narrower
- tall/taller
- thick/thicker
- full/fuller; empty/emptier

POSITION
- up/down; above/below
- towards/away from
- behind/in front of/around
- back/front
- left/right
- top/centre/bottom
- inside/outside
- here/there
- far/near

It is worth noting that some of these describing words included under size are rather ambiguous. For example, how 'wide' is this book? Well, the answer rather depends on whether the book is lying on the table or stacked in a book shelf!

Clearly, you don't need to set up a formal maths activity with your child in order to encourage the use of these sorts of comparing words. They will crop up again and again when you go shopping, cook the dinner or play games like 'Hunt the Thimble'.

Exercise B

Next time you are doing an everyday activity with your child (perhaps packing away the groceries into a cupboard or washing up), see how many of those comparing words you can include naturally in the conversation.

THERE ARE NO COMMENTS ON THIS EXERCISE.

Much of children's play with sand, water and shapes provides them, quite literally, with a 'feel' for quantity and size. This lays down the foundation for more formal measurement that they will tackle later. This is how they begin to discover that they can pour 'small' into 'large' but not the other way round. The only advice to offer here is to encourage such play and encourage your child to talk about what she is doing. For example:

- this one's bigger
- that one holds more
- mine's empty
- pour it into the little one – oh dear, there's too much!

and so on.

Ordering

Getting things in the right order is important for children of all ages. Toddlers are highly amused when you try to put on their socks *after* putting on their shoes.

As they grow older children will use cooking recipes, follow directions, read instructions, and so on. All of these are sequenced activities where the order really matters. A sequence is a particular form of ordering where a number of events are ordered according to

time. The words that will crop up here are 'first', 'second', 'third' . . . 'last', 'before', 'after', 'followed by', 'next', and 'then'. Here are some sequencing activities to try with your child.

Exercise C

Ask your child to:
(i) describe what she does in sequence after she wakes up
(ii) close her eyes and you make a sequence of different noises (for example bell, rattle, clap). Then ask her to say which she heard first/next/last.
(iii) continue to thread beads, following a pattern that you have already started. For example green, red, yellow, green, red, yellow . . .

THERE ARE NO COMMENTS ON THIS EXERCISE.

As well as ordering events over time, we often need to order things by size. For example, it is more efficient to stack the saucepans inside each other before putting them away in the cupboard. When you are tidying away toys the same principle applies. Children's stories often portray their characters in small, medium and large sizes – for example 'The Three Bears' and 'Billy Goats Gruff'. Again you don't need to set up a maths lesson to draw your child's attention to different sizes of household items such as the following:

- bowls and plates
- spoons
- shirts, vests, socks and shoes
- washing powders on the supermarket shelf

The words which your child will use when talking about ordering things in terms of their size will be very similar to those given earlier under the heading 'Comparing'. The difference is that now she is handling more than two items at a time and will be talking about bigg*est*, small*est*, long*est*, as well as first, second, third . . . last.

The set of ten stacking beakers is a fairly standard household toy and one that is fairly difficult for most preschool children. The task can be made simpler if you ask your child to make a row of beakers in order of size and give her just one at a time to put in place. She could also try stacking just three beakers at first. Then gradually increase the number.

Measuring

It is one thing for a child to be able to pick out the biggest and the smallest from a set of beakers and even to stack them inside each other. It is quite another to be able to measure with a ruler the height of a beaker in centimetres and to know its capacity in millilitres. The word 'measuring' is normally taken to refer to this latter activity where size is related to certain conventional units like centimetres or inches. The measure, say of height, is then given as a certain number of these units: 12 cm or 4 inches. The skills involved in this sort of formal measuring (i.e., with conventional units like cm) are quite advanced. For the first few years at primary school your child will probably never talk about centimetres, grams or the like. It is more likely that she will be encouraged to experiment with her own informal (i.e., made-up) measures. For example, length will be measured in matchboxes, paper clips or pencils.

MINE'S BIGGER! LOOKING FOR DIFFERENCES

The skills involved in formal measuring are quite demanding. It is the most refined form of description we have yet met and is beyond the scope of most preschool children. Accordingly the topic is only touched on here, to be taken up again in chapters 14 and 15. However, there is no doubt that your child will absorb some understanding of formal measuring simply by keeping her eyes and ears open. Shopping will expose her to the measurement of money and the units of pounds and pence. Cooking provides the opportunity to see most of the other measures. Have a look at this recipe, for example.

Orange Crusted Jumbles

Makes approximately 30 jumbles
Cooking time: 10–15 minutes   1 egg
350 g (12 oz) self-raising flour   1 orange
100 g (4 oz) butter   1 tablespoon milk
100 g (4 oz) sugar   1 egg white

Grate the rind of the orange and squeeze out the juice. Cream the butter and sugar until white. Add the egg, orange rind and juice. Fold in the sieved flour and use the milk to mix to a firm dough.

Roll out the dough 5 mm (¼ inch) thick and cut into narrow strips about 15 cm (6 inches) long. Coil each strip like a flat Catherine wheel. Place the coils on a greased baking sheet, brush with white of egg and sprinkle with nibbed sugar (sugar crystals that are used, for example, to decorate bath buns). Cook at 230°C (450°F), gas mark 8, for 10–15 minutes.

From *Reader's Digest Book of Family Games and Party Treats* (1979)

Exercise D

(i) Mark on the recipe for Orange Crusted Jumbles four references to measurement that you can find.
(ii) Now complete the table below indicating for each word the unit of measure and the thing being measured (the first one has been done for you to get you started).
(iii) Now get your apron on, round up the nearest toddler(s) and get stuck into the Orange Crusted Jumbles – mmm, delicious, and so educational!

Example	Unit	Measure
1. Cooking time: 10–15 minutes	minutes	time
2.		
3.		
4.		

COMMENTS ON THIS EXERCISE ARE GIVEN ON PAGE 63.

Conclusion

The main message of these chapters in the preschool section has been that maths, even for toddlers, is helpful in making clear descriptions of the world around us. Chapter 3 looked at the way we make sense of this world by looking for similarities in the objects and events we see. This idea was mirrored in chapter 5 where the main theme was looking for differences. Between these came chapter 4 which listed the skills which a child must acquire before she can count accurately. The skills involved in looking for similarities (chapters 3 and 4) were matching, sorting, classifying and counting. Looking for differences (chapter 5) highlights the skills of comparing, ordering and measuring.

It should be stressed that whatever activity your child may be engaged in – playing, shopping, cooking – it is likely that all of these skills will be in evidence to some degree. The distinctions between them are not clear cut and there is not always an obvious sequence of development for a given child.

Some games and activities with a preschool child

Long journeys and wet weekends can be pretty tedious for both parents and children. Here are a few ideas to keep them going.

1. *Jigsaws*
You can make your own jigsaw by gluing a suitable picture to cardboard and cutting it into a few pieces. For young children the cuts should be kept simple and each piece should contain a recognizable part of the picture.

2. *Rhythms*
Tap out a simple rhythm on the child's arms, legs or body and then ask her to beat it back to you from memory. Gradually make the rhythms more complicated. Later the rhythms can be clapped or tapped on a table. It is helpful to make up nonsense words which convey the rhythm, like 'Tum tumty tum tum'.

Try clapping even rhythms where the accent varies. For example, in three time:

X X X | X X X | X X X | . . .

or in two time:

X X | X X | X X | . . .

and so on.

If your child gets good at spotting rhythms, try beating out the rhythm of a well-known tune (say 'Baa Baa Black Sheep') and ask her to identify the song. Then she can do one for you. This last game may be beyond most five-year-olds but by the time they are six or seven, many children could tackle this with much enjoyment.

3. *I Spy*

I Spy can be played almost anywhere – in the kitchen, on a long journey or reading a book together. Try playing it as follows:

I Spy – a square
– a circle
– a rectangle
– three things together
– a pair

4. *Gestures*

Make up various postures using your arms, legs and body. Get your child to copy them precisely. Gradually make the gestures more and more subtle.

Now try a sequence of two, then three gestures to be copied by your child. Then get her to initiate and you follow.

5. *Treasure hunt*

Hide a number of objects round the house. The clues can be pictures or maps – later they can be written instructions or cryptic clues. In the interests of economy each hidden object can contain a note which provides the location of a grand prize. (This is how Christmas presents are distributed in our house and it is considered to be an essential part of the spirit of Christmas. It certainly taxes the old brain late on Christmas Eve thinking out the cryptic clues!)

MINE'S BIGGER! LOOKING FOR DIFFERENCES

ANSWERS TO EXERCISES FOR CHAPTER 5

Exercise A
No comment.

Exercise B
No comment.

Exercise C
No comment.

Exercise D

Example	Unit	Measure
1. Cooking time: 10–15 minutes	minutes	time
2. 350 g (12 oz) self-raising flour	grams/ ounces	weight
3. 1 tablespoon milk	tablespoon	capacity
4. Roll out the dough 5 mm (¼ inch) thick	mm/inches	length (or depth or height)
5. Cook at 230°C (450°F), gas mark 8	°C/°F/ gas mark 8	temperature

Part II

The Primary Years

6
What maths is all about

Part I looked at some of the most basic ideas in maths like sorting and comparing – so basic that even very young children can't avoid using them as they try to make sense of the world. Part II will focus on the more usual topics which comprise school maths and which your child will almost certainly cover between the ages of five and twelve years. An important feature of these chapters is that you will be expected to do some maths yourself, so arm yourself now with a pencil, paper and calculator. The cheapest calculator you can find will be suitable and if you don't already own one you can probably pick one up from your local petrol station for about £2. You might be wondering 'Why a calculator?' The answer is that teachers are increasingly discovering the possibilities of learning maths through a calculator. They can see how it helps to motivate their students and build the confidence needed to tackle questions that children wouldn't otherwise attempt. However, it should be stressed that although the calculator is a substitute for *calculation,* it is not a substitute for thinking. The user still needs to decide what keys to press and to be able to interpret the answer in the calculator display. Some suggestions for using and exploring your calculator are given in chapter 22.

What is maths?

Just what can you remember about the maths you learnt at primary or secondary school? Go on – get out a pencil and paper and jot down some of the words which cropped up in maths lessons.
 Your list might look something like this:

arithmetic, decimals, equations, *geometry*, addition, tables, fractions, circles, square roots, *algebra* . . .

Most of the words in the list refer to maths topics (decimals, equations, fractions . . .). However, the three which are in italics are whole branches of mathematics. If you are like the Bradford mothers, you won't be able to sort out which is which. One of them, Dorothy, once commented, 'It's like a Scotch mist to look at, never mind do!'

We'll make a start by trying to piece together some of the important topics of maths so that the complete jigsaw fits together. As part II progresses, each new topic will be slotted into position so that you can see where it belongs in the overall picture.

The traditional way of thinking about school maths has been in terms of the three major branches, arithmetic, geometry and algebra. Arithmetic involves calculating with numbers, geometry concerns shapes and patterns, while algebra uses letters like x and y to express relationships between things. Here is a *tree diagram* showing how just some of the more common topics fit into this way of looking at maths.

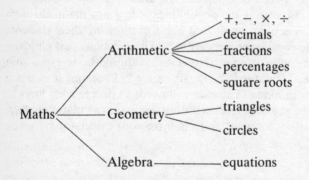

This picture of maths might be fairly familiar to most of you. But what about this 'new maths' that children are doing now? Well, don't panic, because it isn't very different from what you see here. There are three main reasons why it all looks so new. Firstly, teachers tend not to talk about arithmetic, algebra and geometry any more. The reason is that there is no longer a clear distinction.

Twenty years ago they tended to be thought of as separate subjects like this:

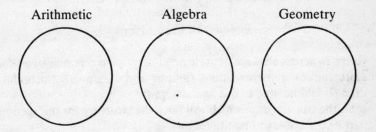

But more recently teachers have tried to encourage children to connect up these separate ways of looking at maths and the picture is more like this:

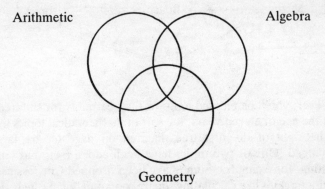

A second change has been that school maths is now more concerned with helping children to understand concepts, rather than practising mechanical skills like long division. Sometimes, however, parents feel that schools go to the other extreme ('It's all very well him knowing that 2×3 gives the same answer as 3×2, but when is he going to learn that they both give 6?')! Clearly a balance needs to be struck between understanding concepts and practising skills. Finally, teachers seem to be more interested in finding practical and realistic applications for maths than used to be the case. Increasingly we have the calculator to thank for this healthy development.

The very words 'arithmetic', 'geometry' and 'algebra' don't provide much of a clue as to the sort of maths they involve. In this text they are replaced with the more descriptive terms:

numbers, *space* and *logic*

Running across all three of these headings we have a range of useful mathematical representations (graphs and diagrams) which children should be able to use and interpret.

So the tree diagram which will form the structure for this second part of the book will be the following:

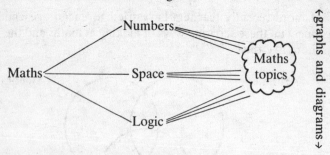

However, you'll be pleased to know that, certainly for children up until the age of eleven years, the sort of mathematical topics listed on the right of the first tree diagram on page 68 are largely unchanged. One or two topics have been added (sets and binary counting, for example). Some have been dropped (the less useful arithmetic skills like calculating square roots and, increasingly, long division). Despite what you may hear to the contrary, good old-fashioned arithmetic is still the main course in the maths diet offered by most primary schools, and the main innovation has been to try to make this diet more palatable and relevant.

Here now are one or two questions that the Bradford mothers asked me and may have been concerning you also.

Why does maths look so complicated?

It is not the ideas in maths that are hard to understand so much as the language and symbols we use. But though these may make it hard to begin with, they have certain advantages over ordinary English. In particular, maths language is:

concise, so we can express a complex idea in a short mathematical 'sentence'. This is achieved by using certain conventions which have a specific meaning but which may not be obvious to a non-mathematical reader.

For example, my cookery book contains what it calls a 'Quick Guide to Roasting Time'. This contains several charts showing suggested times for different oven temperatures, and the nature and weight of meat. Although most people find these charts helpful and easy to use, much of the information can be reduced to one or two simple formulae. Here is a formula for roasting lamb.

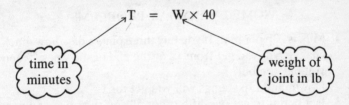

What this formula tells us is that the roasting time for any joint of lamb can be found by multiplying its weight in lb (W) by 40. In other words, you allow 40 minutes per lb.

What is the point of learning maths?

Just like being able to read, there are certain basic maths skills that you need in order to live a normal life. For example, being able to:

- read numbers and count
- tell the time
- handle money when shopping
- weigh and measure
- understand timetables and simple graphs.

But as well as having a useful, practical side, solving maths problems can also be challenging and fun. Anyone who has done dressmaking or carpentry knows that maths can be used in two ways. One is in the practical sense of measuring and using patterns and diagrams. The other is more abstract – 'How can I cut out my pattern so as to use up the least material?', 'Can I use the symmetry of the garment to make two cuttings in one?' The pleasure you get from solving *these* sort of problems is what has kept mathematicians going for four thousand years!

Most of all, mathematics is a powerful tool for expressing and communicating ideas. Sadly, few children (or adults) ever get a sense of the 'power to explain' that mathematics offers.

So much for talking about maths. Now it is time to *do* some sums.

PRACTICE EXERCISES

SOME EVERYDAY PROBLEMS

1. Milk is 23p per pint. If you buy three pints a day, how much change will you get from £5 at the end of the week, when you pay the bill?
2. How many 17p stamps will you get for £1?
3. Is it better to get one-fifth off or 5% off the price of a new cooker?
4. How much is my electricity bill (fill in the boxes)?

METER READINGS

Present	Previous	Units supplied	Pence per unit	Amount (£)
56769	56242		5.10p	
			Fixed Charges	£4.64
			Total	

ANSWERS ARE GIVEN ON PAGE 269.

7
Whole numbers

What is a number?

Have you ever thought about what the number 3 means? Well one meaning is as a way of describing a *set* of things. Here is a set of dots.

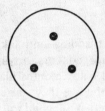

And this is a set of calculators:

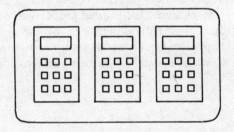

Both these sets have something in common – yes, you've guessed it – their 'threeness'. It is a major breakthrough for your child when she learns that 'three' is a description that can be applied to a whole variety of different collections (or sets) of things. She will need many occasions in which she physically handles three objects before threeness becomes a concept that your child can grasp. This is the *cardinal* property of number described in chapter 4 and which your child's teacher will undoubtedly seek to establish with the use of structured apparatus (for example, Cuisinaire rods, Dienes blocks, etc.).

Also mentioned in chapter 4 was the *ordinal* property of numbers. This refers to the way that they follow on from each other in sequence 1, 2, 3, 4, 5 . . . The *number line* shown below is a helpful way of enabling children to form a mental picture of the sequence of numbers.

```
———————————————  A NUMBER LINE
 0  1  2  3  4  5
```

Children first learn about 'adding' and 'taking away' with physical objects (bricks, bottle tops, conkers, or whatever is available). In infant school they spend many months playing games like taking two marbles away from the five they started with and finding that they have three marbles left. Your child will be able to do this sort of subtraction with marbles, sweets and conkers long before she is able to grasp the abstract number fact that '5 take away 2 gives 3'. Later she can start to think about the *ordinal* property of numbers. For example, children of seven and eight can use the number line to 'picture' what happens when they add or subtract two numbers. The diagram below shows the number line being used to find the answer to:

$$5 - 2 =$$

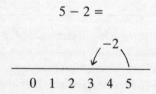

You simply start at 5 and move 2 steps to the left. (For addition, you move to the right.)

WHOLE NUMBERS

What are the 'four rules'?

The four most basic things you can do to any two numbers is add (+), subtract (−), multiply (×) or divide (÷) them. These are the four rules (sometimes called the four 'operations'). What causes some confusion for children is that different teachers use different words to describe them. How many words can you think of which refer to the four rules? Most of them are listed in Exercise A where you'll get a chance to see how many you recognize.

Exercise A   The words used to describe the four rules

Tick which word refers to which rule.

Word	Rule (+ − × ÷)	Word	Rule (+ − × ÷)
add	✓ (+)	multiply	✓ (×)
and	✓ (+)	plus	✓ (+)
difference	✓ (−)	product	✓ (×)
divide	✓ (÷)	share	✓ (÷)
goes into	✓ (÷)	sum	✓ (+)
how many more	✓ (−)	subtract	✓ (−)
how many less	✓ (−)	take away	✓ (−)
minus	✓ (−)	times	✓ (×)

COMMENTS ON THIS EXERCISE ARE GIVEN ON PAGE 86.

HELP YOUR CHILD WITH MATHS

How do the four rules relate to each other?

The four rules are, of course, closely connected to each other. When this particular penny drops for your child she will find that her grasp of number patterns quickly improves. Twenty minutes 'playing' with numbers on a calculator can lead her (with a little help from you) to reach conclusions like:

$\boxed{+}$ is the opposite of $\boxed{-}$ (they go in opposite directions on the number line)
$\boxed{\times}$ is the opposite of $\boxed{\div}$
$\boxed{\times}$ is a quick way of doing lots of $\boxed{+}$s
$\boxed{\div}$ is a quick way of doing lots of $\boxed{-}$s

Exercise B is designed to let you 'discover' for yourself some of these conclusions about numbers. It is best done with the help of a calculator, though this is not essential.

Exercise B   Connecting the four rules

Fill in the blanks in the sequences below.
(Those blanks marked ◯ refer to a number. Blanks marked ☐ refer to a 'rule' like + or −.)
  Then jot down on the right-hand side what you think this suggests about the four rules.

Key sequence	Comment
3 $\boxed{+}$ 2 $\boxed{=}$ ⑤ $\boxed{-}$ 2 $\boxed{=}$ ③	Brill
4 $\boxed{+}$ 3 = ⑦ $\boxed{-}$ 3 $\boxed{=}$ ④	Brill

WHOLE NUMBERS

Key sequence		Comment
3 × 2 = ○ ÷ 2 = ○		
4 × 3 = ○ □ 3 = ④		
20 ÷ 5 = ○ □ 5 = ⑳		

0 + 3 + 3 + 3 + 3 = ○	How many times did you add 3? □
4 × 3 = ○	

6 − 2 − 2 − 2 = ○	How many times did you subtract 2? □
6 ÷ 2 = ○	

COMMENTS ON THIS EXERCISE ARE GIVEN ON PAGE 86.

What is 'place value'?

'Place value' is the posh name for what we used to call 'hundreds, tens and units'. You can't make any sense of numbers larger than nine unless you understand place value. For example, the number 33 has two *digits*, both of which are 3. The number 444 has three *digits*, all of them 4. But their *position* (hence the word 'place') in the number tells us whether they represent 4 units, 4 tens or whatever. The calculator game given on page 267, called 'Space Invaders', is designed to help children with place value. If you have trouble with decimals, this game will get you off to a good start.

5 from 3? You can't... can you?

You may have noticed that the subtraction sums which your ten-year-old does at school are always 'fixed' so that she is taking smaller from larger. Young children see subtraction very much in terms of 'taking away' objects and clearly if you start with three bricks you can't take more than three away. However, subtraction doesn't always involve moving objects around. Look at these two examples:

With £3 in my bank account I wrote a cheque for £5.

The temperature was 3°C and it dropped a further five degrees overnight.

The banking system hasn't ground to a halt or the thermometer exploded as a result of these events. We simply solve the problem by inventing a new set of numbers less than 0. In bank statements these have the letters O/D (standing for overdrawn) beside them. Usually, however, we just call them *minus* numbers or *negative* numbers. So really the number line should be extended to look like this.

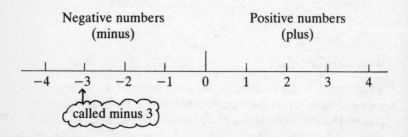

Subtracting 5 from 3 can be shown on the number line as follows:

WHOLE NUMBERS

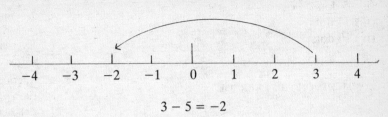

$$3 - 5 = -2$$

Some special numbers

> The Pooka MacPhellimey, a member of the devil class, sat in his hut in the middle of a firwood meditating on the nature of the numerals and segregating in his mind the odd ones from the even.
> From *At Swim-Two-Birds* by Flann O'Brian

As a member of the human class, I expect you feel that you spend your meditating time more imaginatively than the Pooka MacPhellimey. However, human or not, it is quite fun looking at the whole numbers from 0 to 100 and seeing what properties they have. For example, try Exercise C now.

Exercise C

Six dots can be arranged in a rectangle like this

. .
. . , or this . . .
. .   . . .

See if you can arrange the following number of dots into rectangles.

(i) 9 dots
(ii) 5 dots
(iii) 11 dots
(iv) 10 dots

COMMENTS ON THIS EXERCISE

As you may have discovered, some numbers (like 5 and 11) cannot be arranged in a rectangle. These are called *prime* numbers. This means that the only numbers which will divide exactly into them are themselves and 1. *Rectangular* numbers (sometimes known as *composite* numbers) have *factors* other than themselves and 1. (The factors of 6 are 3 and 2 because 3 *times* 2 = 6.) Of the composite numbers, some like 4, 9, 16 . . . are particularly special. For obvious reasons these are called *square* numbers. Finally, there are the *odd* numbers (1, 3, 5, 7 . . .) and the *even* numbers (0, 2, 4, 6 . . .). These are so called because if you think of sharing a certain number of sweets between two people it will either divide 'evenly' or else there will be an 'odd' one over.

The final exercise in this chapter asks you to think about these special numbers and see if you can come up with some of their general properties. Choose some particular numbers to get you started.

Exercise D

(i) Does the square of an even number always give an even number? _yes_

(ii) Does the square of an odd number always give an odd number? _no_

(iii) Are all odd numbers prime? _no_

(iv) Are all prime numbers odd? _yes_

WHOLE NUMBERS

(v) Write down the numbers from 1–20 and, using a table like the one shown below, tick whether each number is prime, composite, odd, even or square.

Number	Prime or composite	Odd or even	Square
1	✓	✓	✓
2	✓	✓	✓
etc.			

COMMENTS ON THIS EXERCISE ARE GIVEN ON PAGE 87.

Where whole numbers fit into maths

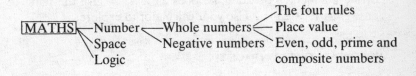

PRACTICE EXERCISES
(Try to spend about thirty minutes on these exercises.)

1. Set yourself a few simple 'sums' using the four rules of +, −, × and ÷. Check your answers using a calculator.
   Do the 'sums' involving + and − again by drawing a number line and moving, respectively, right or left. Check that you get the same answer as with the calculator.

2. *Place value*
   The place value of the 6 in the number 365 is ten.
   What is the place value of the 6 in the following numbers?

Number	365	614	496	16042	1093461
Place value	10				

3. Play Calculator Space Invaders levels 1 and 2 (see chapter 22).

4. 7°C is three degrees less than 10°C.
   Find the temperature which is three degrees less than the following:

Temperature °C	10	4	21	−6	−10	0	3	−3
Three degrees less	7							

5. Does it matter in what order you add, subtract, multiply and divide numbers? For example, does 23 × 15 = give the same answer as 15 × 23 = ? Use your calculator to explore.

ANSWERS ARE GIVEN ON PAGE 270.

Helping your child understand whole numbers

Most of the ideas contained in the preschool section (chapters 2–5) lay the foundations for your child's understanding of number. Many of the counting suggestions listed in chapter 4 are relevant to older children of six, seven and eight years. It is worth stressing, however, that children find it difficult to think of number in the abstract. They need to handle three objects and actually *feel* their threeness. They also should be encouraged to recognize that number is both a useful and a precise way of describing things. Numbers provide us with the basic data of our existence. Here are some examples of basic cardinal and ordinal number facts which young children should get to know about themselves. (If you've forgotten what cardinal and ordinal numbers are, see page 43.)

Cardinal number facts

- the number of people in my family
- the number of children in my class
- money – the cost of various commodities
- how many ears, noses, fingers and toes I have

Cardinal numbers will crop up all the time and it won't seem odd to your child to draw attention to them. For example, she might help you discover how many fish fingers or sausages each member of the family wants to eat and place the appropriate number on the grill pan.

Ordinal number facts

- house number and telephone number
- age of members of the family
- size of shoes and clothes

DO NOT try to 'teach' your child the difference between cardinal and ordinal numbers. It is enough simply to use numbers when they help to describe things better. Give your child lots of practice using *objects* to represent numbers before moving on to *symbols* (numerals) or a number line. As with most maths learning, understanding about number is built up in the following stages:

Get your child *talking* about what she is doing at each stage and try to support her understanding throughout with the calculator.

Board games, playing cards and dice games are good family entertainment for a wet weekend and all help to foster a friendly acquaintance with numbers. Learning to tell the time and playing with a calculator will help your child to recognize the numerals and reinforce in her mind the order in which they come. Best of all, you can't beat a good story, and it is surprising how often numbers are a key feature of popular children's stories. Using a combination of pictures and words it is easy to convey the threeness of the little pigs, the blind mice and Billy Goats Gruff, or the seven-ness of Snow White's dwarfs.

So much for using numbers with young children. Older children will be getting to grips with symbols and the four rules of number ($+$, $-$, $\times$ and $\div$) and need as many everyday examples of these rules as you can provide. The *taking away* and *adding* of objects need to be done many times physically before children can grasp the concepts of subtraction and addition as abstract operations. The same sort of acting-out needs to be repeated for multiplication (repeatedly *adding a number of times*) and division (both *taking away a number of times* as well as *sharing among a number of people*). So whether you are cooking, serving out meals or sharing a bar of chocolate, there is much that your child can be encouraged to do and say to develop her grasp of the four rules of number.

And now to the vexed question of number tables! Many parents are concerned when their children don't seem to know their tables properly. Most of today's parents probably learnt their times tables by drill in the primary school and, although it was probably stressful at the time, feel that their children aren't doing real sums if they aren't proficient at chanting tables. After all, what was bad enough for us must be good enough for them! Well, there is no 'correct' view on this question. My own feeling is that having a good grasp of number bonds (i.e., knowing your tables) is a useful skill and one which I would like my children to acquire. However, I'm not convinced that this is best achieved learning tables by drill and I don't think it should be forced on a child too soon. I don't see it as my job as a parent to make the learning of tables another area of stress between me and my child (it's as much as I can cope with 'encouraging' their help with the washing up and keeping their bedrooms tidy!). What a parent *can* do is to build confidence with number and encourage curiosity about patterns in numbers. There are several calculator activities and games given in chapter 22 which you can play with your child and which will help her to develop number sense. Some further games are given in the next chapter. Most of all, however, try to involve your child in family decision-making. Numbers and number operations crop up everywhere – in work, leisure, shopping, the home and in the general life of the community. Seeing numbers being used in a wide variety of situations will give your child a better understanding of *when* to add, multiply, and so on, as well as how to make sense of her answer in a practical problem. These important questions are explored again in chapter 13.

Below is a checklist which indicates the main number skills which your child should be aiming to acquire. Please note that your child won't be able to demonstrate competence in all of these skills so *don't expect too much!*

Number

The ability to:

- count a collection of objects
- count out a number from a larger number (e.g., peel me five grapes)
- count on (e.g., adding 3 to 4; count 5, 6, 7)
  (try this on the calculator)
- make up the 'story' of any given number (for example, the story of 6; $3 + 3$, $4 + 2$, 3×2, etc.)
- recognize the numerals
  (again, exploit the calculator)
- know that 5 is greater than 3 and 4 is less than 7
- understand place value (hundreds, tens and units)
  (play Calculator Space Invaders – see page 267)
- recognize odd, square, even and prime numbers
- be aware of negative numbers

Number operations (+, −, × and ÷)

- understand the concept of the 'four rules'
- use and understand the vocabulary of the 'four rules' (e.g., share, times, product, sum, difference . . .)
- know that + and × give the same answer when done in any order but − and ÷ don't
- realize that multiplication can be thought of as repeated addition
- realize that division can be thought of as repeated subtraction or as sharing
- make reasonable estimates of the approximate answer to a calculation
- know what 'sum' to do in a practical situation

ANSWERS TO EXERCISES FOR CHAPTER 7

Exercise A

WORD	Rule + − × ÷					WORD	Rule + − × ÷			
add	✓					multiply			✓	
and	✓					plus	✓			
difference		✓				product			✓	
divide				✓		share				✓
goes into				✓		sum	✓			
how many more		✓				subtract		✓		
how many less		✓				take away		✓		
minus		✓				times			✓	

Exercise B

This was designed to get you thinking about the connections between the four rules. These were:

Number	Prime	Composite	Odd	Even	Square
10		√		√	
11	√		√		
12		√		√	
13	√		√		
14		√		√	
15		√	√		
16		√		√	√
17	√		√		
18		√		√	
19	√		√		
20		√		√	

WHOLE NUMBERS

(i) ⊞ is the opposite of ⊟
(ii) ⊠ is the opposite of ⊡
(iii) ⊠ can be thought of as repeated adding
(iv) ⊡ can be thought of as repeated subtracting

Exercise C

No comment.

Exercise D

(i) Yes.
(ii) Yes.
(iii) No. There are many exceptions like 9, 15, 49 . . .
(iv) All prime numbers are odd except 2.
(v)

Number	Prime	Composite	Odd	Even	Square
1	√		√		√
2	√			√	
3	√		√		
4		√		√	√
5	√		√		
6		√		√	
7	√		√		
8		√		√	
9		√	√		√

8
Some games and activities with numbers

Long journeys or wet weekends at home can be boring for both parents and children. This chapter provides a few suggestions for some number puzzles and activities which should help to amuse and entertain.

1. *Number Plate Games*

   Games with car number plates can be played anywhere near a road or car park. See if you can spot numbers where:
   (i) the digits add to 10
       (e.g., A163 VVV)
   (ii) all the digits are even
       (e.g., B426 EJI)
   (iii) all the digits are odd
       (e.g., FLO 139W)
   (iv) there are two digits which differ by 1
       (e.g., FOR 384Y)
   (v) all of the digits are prime
       (e.g., MAT 715W)
       etc.

2. *Pub Cricket*

   Players take turns to 'bat'. You score from the number of legs (human or animal) which can be seen on each pub sign you pass. For example, the 'Bull and Butcher' scores six, while the 'White Hart' scores four. If you go past a pub sign which has no legs then you are out and the next player takes a turn at batting.

Popular signs in this game, by the way, are the 'Coach and Horses' and the 'Cricketers' Arms'!

3. *Guess My Number*

One player picks a number between 1 and 100 and the other player must guess it with as few questions as possible. Note that the questions must be such that they require yes/no answers. Useful questions are ones such as:

'Is it less than 50?

or 'Is it even?'

Less useful questions are ones such as:

'Is it 26?'

since this eliminates only one number at a time.

4. *The Story of 12*

How many *different* stories can you make of 12?
Well, it is $11 + 1$, $10 + 2$, etc.
Don't forget it is also $12 + 0$.
What about 4×3, 3×4 and 6×2.
And don't forget 12×1.
Now we're getting stuck. Ah! it's 24 lots of a half, so it's $24 \times \frac{1}{2}$.
Well, if you're allowing fractions it's $1\frac{1}{2} + 10\frac{1}{2}$.
Hey, this could go on all night!

5. *Finger Tables*

Most people know their 'times table' up to about 5. However, with 6 and above they may have problems. Here is a method which provides the answer to all products between 6 and 10, using the cheapest digital calculator around – your fingers.

Place your hands in front of you as in the diagram, thumbs uppermost.

Starting with the thumb, number the fingers as shown.

Now to multiply, say, 7 and 8, touch the 7 finger of one hand with the 8 finger of the other (it doesn't matter which way round).

Now the answer to 8 × 7 can be found as follows:
(a) Count the number of fingers *below and including* the touching fingers (in this case 5). This gives the number of tens in the answer.
(b) Multiply the number of fingers on each hand *above* the touching fingers (here it is 3 × 2 = 6). This gives the number of units in the answer.

6. *Magic Squares*

This is a magic square.

8	1	6
3	5	7
4	9	2

It is so named because all the rows, columns and diagonals 'magically' add to the same total (15). Can you make a 4 × 4 magic square? Or a 5 × 5?

Hint: You may have spotted that the total of 15 for the 3 × 3 magic square is three times the middle value of 5. Can you first of all work out what the rows, columns and diagonals of the 4 × 4 square should add to?

7. *Magic Triangle*

The numbers 1, 2 and 3 have been placed at the vertices (i.e., the corners) of the triangle.

Can you place the numbers 4, 5, 6, 7, 8 and 9 along the sides of the triangle so that the four numbers on each side (i.e., including the numbers at the vertices) add up to 17?

8. *Upside Down*

The year 1961 reads the same when turned upside down. When was the last year before 1961 that reads the same upside down? What digits become letters of the alphabet when turned upside down? What calculation on your calculator would produce the following words when turned upside down?
(i) hELLO
(ii) OhELL
(iii) ELSIE
See if you can make up a calculator crossword puzzle using calculations as clues whose answers, upside down, fit the puzzle.

9. *Four 4s*

The number 7 can be expressed using four 4s as follows

$4 + 4 - 4/4$

Express the numbers 0, 1, 2 . . . 10 using four 4s and any other non-number keys on the calculator.
Hint: 2 can be written as $\sqrt{4}$.

10. *The Bells, the Bells*

(a) If it takes 15 seconds for a church bell to chime 6 o'clock, how long does it take to chime midnight?

(b) A fence panel is 6' long. How many fence posts are needed to panel a 48' gap?

(c) What have (a) and (b) in common? Can you think up any more questions of your own based on the 'posts and spaces' problem?

SOME GAMES AND ACTIVITIES WITH NUMBERS 93

11. *Return Journey*

I plan to complete a return journey (there and back) in an average time of 40 mph. However, my outward journey is slow and I complete that part in 20 mph. How fast must I travel on the return journey to average 40 mph overall?

12. *Calculator Games*

Several calculator games are described in chapter 22. You might like to try some of these with your child.

If you enjoy these sort of puzzles, there are a number of entertaining collections available. See references 2 and 3 (page 277).

ANSWERS TO EXERCISES FOR CHAPTER 8

Exercises 1–5

No comments.

Exercise 6

With this magic square each row, column and diagonal adds to 34. What makes it even more magic is that every block of 4 squares also add to 34!

7	12	1	14
2	13	8	11
16	3	10	5
9	6	15	4

Exercise 7

Exercise 8

1881; ¾ + 0.0234; 65 × 1000 + 12340; 33333 − 1760.

Exercise 9

$0 = 4 + 4 - 4 - 4$ or $\frac{4-4}{44}$

$1 = 4/4 + 4 - 4$ or $44/44$

$2 = 4/4 + 4/4$

$3 = 4/4 + 4/\sqrt{4}$ or $\frac{4+4+4}{4}$

$4 = 4 + \frac{4-4}{4}$

$5 = \sqrt{4} + \sqrt{4} + 4/4$

$6 = \sqrt{4} + \sqrt{4} + 4/\sqrt{4}$

$7 = 4 + 4 - 4/4$ or $44/4 - 4$

$8 = 4 + 4 + 4 - 4$

$9 = 4 + 4 + 4/4$

$10 = 4 + 4 + 4/\sqrt{4}$

Exercise 10

(a) 33 seconds (for 6 chimes there are 5 intervals. Therefore it must take 3 seconds per interval. Twelve chimes has 11 intervals, hence 33 seconds).

(b) 9 posts are needed for 8 spaces.

(c) Both the above questions refer to a common 'type' of maths question known as the old 'posts and spaces' trick.

Exercise 11

It can't be done! Suppose the total distance (there and back) is 40 miles, then the total journey (there and back) must take exactly one hour. If the outward journey of 20 miles is completed at a speed of 20 mph, the one hour is completely used up!

Exercise 12

No comment.

9
Fractions

> Decimals, now, and fractions. I don't know, I just didn't seem to grasp them . . . I just found them boring. I couldn't concentrate on them at all. They weren't interesting enough for me. It wasn't attractive enough.
>
> A Mum

Not the most promising start to a lesson on fractions perhaps! It certainly seems to be the case that while most people know roughly what's going on when the four rules are applied to *whole* numbers, sums with fractions are 'like a Scotch mist'. The first thing you should realize is that fractions, decimals and percentages are all very similar. As you will see over the next three chapters, they are all slightly different ways of describing the same thing. What we call fractions – things like ½, ¾ and so on – should really be called *common* fractions. In fact these sort of fractions are not quite as 'common' as they used to be. Increasingly the more awkward common fractions are being replaced by *decimal* fractions. Decimal fractions look like 0.3, 0.125, and so on and are dealt with in chapter 10. But first of all, let's find out what a fraction is and where it comes from.

What is a fraction?

Emma and Sam are four. They have never heard of a fraction. I produced three squares of chocolate and said that they were to be shared between them. They took one square each. Now what about that third square? Well, you can be sure that they won't give it to

me, or their favourite charity. Emma and Sam may not have heard of a fraction, but they are quite capable of inventing one when the occasion arises.

Fractions can be thought of as the 'broken bits' that lie between the whole numbers. Fractions occur quite naturally in division (i.e., sharing) when the sum doesn't divide exactly. For example, sharing seven doughnuts amongst 3.

$7 \div 3 = 2$ remainder 1

But as with the square of chocolate, we don't always want to leave the remainder 'unshared'. If the remainder of 1 is *also* shared out amongst the 3 (people) they each get an extra one third. So, the more complete answer to this division sum is:

$7 \div 3 = 2\frac{1}{3}$

It is important that children understand why fractions are written as they are. $\frac{1}{3}$ really is another way of writing $1 \div 3$. So the top number in a fraction is the number of things to be shared out. The bottom number tells you how many shares there will be.

FRACTIONS

Exercise A will help you grasp this important idea.

Exercise A

A box of processed cheese has six segments.
Share two boxes equally among three children.

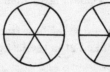

(i) How many segments will each child get? ____6____

(ii) How many segments are in each box? ____6____

(iii) What fraction of a box will each child receive? ____½____

COMMENTS ON THIS EXERCISE ARE GIVEN ON PAGE 106.

How to picture a fraction

The common fractions like a half, three quarters and two thirds are part of everyday language. Your child will learn those words from you as a matter of course, in the same way as she learnt much of her vocabulary up till the age of five. By the age of six or seven she will find it helpful to have a mental picture of a fraction. The picture which is in my mind (and is used in most schools) is to image a *whole* as a complete *cake*. This can be cut into slices representing various fractions, like this:

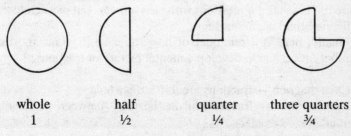

whole   half   quarter   three quarters
1   ½   ¼   ¾

Talking about cakes helps children to understand what a fraction is. But before long, children find that they need to compare fractions.

Is ⅔ of a cake bigger than ¾ of it?

Children need to see how fractions fit into the sequence of numbers that we looked at in the last chapter. For example:

3⅓ is one third of the way between 3 and 4

5³⁄₁₀ is three tenths of the way between 5 and 6

The best way to get this message across is to find where the fractions fit on the number line.

A number line showing fractions

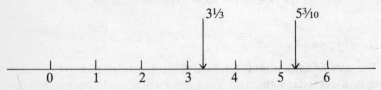

This diagram is a considerable help in teaching about numbers to children of all ages. Until fractions are introduced, numbers can be thought of as a set of points equally spaced on the line (i.e., the whole numbers). Suddenly, children begin to see that there are lots of other numbers between 1 and 2; 2 and 3. How many are there? Are there any gaps at all on the number line when the fractions are added? These aren't questions with easy answers but most children will enjoy thinking about them.

Finally, here is a reminder of how the cake diagram and the number line can help develop a mental picture of fractions.

Cake diagram → fractions are 'bits' of a whole
Number line → fractions fill in the gaps 'between the whole numbers'

And now you're ready to add and subtract fractions. Well . . . nearly. . . . Before that it would be useful to know what equivalent fractions are.

What are equivalent fractions?

What is the difference between sharing two cakes among four people or sharing one cake between two people? Well, since everybody ends up with half a cake, there is no difference. The first lot of people actually get $2/4$ of a cake but that would seem to be the same as $1/2$.

So $2/4$ and $1/2$ are fractions which are the same, but different – they have the same value but have different numerals top and bottom.

The word used to decribe this is *equivalence*. We would say that $1/2$ and $2/4$ are *equivalent fractions*. Exercise B(ii) will give you a chance to try to spot some more equivalent fractions. Try Exercise B now.

Exercise B

(i) Mark with an arrow these numbers on the number line below: $2\frac{1}{2}$, $3/4$, $4\frac{9}{10}$, $1\frac{1}{3}$

```
|____|____|____|____|____|
0    1    2    3    4    5
```

(ii) Find three fractions equivalent to each of the following:

Fraction	Equivalent fractions
$3/4$	$6/8$, $9/12$, $12/16$
$6/18$	
$2/5$	
$10/20$	

COMMENTS ON THIS EXERCISE ARE GIVEN ON PAGE 106.

How do I add and subtract fractions?

When adding fractions, it is helpful to think of the slices of cake. For example:

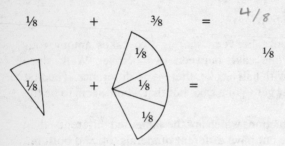

Here the slices are all the same size (⅛ each) so we just add them together, like this:

⅛ + ⅜ = 4/8

It is usual to write answers in the form of the simplest equivalent fraction, so the answer = ½. However, what happens when you have to add fractions like

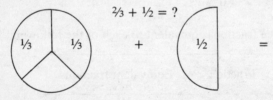

This time the slices of cake aren't the same size so we can't just add them together. The way out of this problem is to cut both fractions up into smaller slices until *all* the slices are the same size – like this:

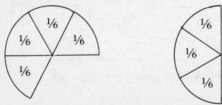

FRACTIONS

Now, with all the slices equal to ⅙, they *can* be added. The sum looks like this:

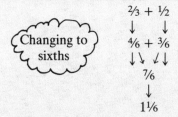

But why did I choose to subdivide each fraction into slices of ⅙? The reason is that ⅙ is the easiest fraction that a half and a third will break up into. I could have used slices of ¹⁄₁₂ or ¹⁄₁₈ but that would have been unnecessarily complicated.

By the way, this process of breaking fractions up into smaller slices so that they can be added or subtracted is called 'finding the smallest common denominator'. Does that phrase come wafting back from your early school days? Let me remind you that finding lowest common denominator means finding the smallest number which both the denominators will divide into. This number then becomes the new denominator. Thus in the example above the lowest number that 3 and 2 both divide into is 6, so 6 is the new denominator.

Now have a bash at Exercise C.

Exercise C   Adding and subtracting fractions

Complete the table below:

Sum	Equivalent fractions	Answer
⅔ + ¼	⁸⁄₁₂ + ³⁄₁₂	¹¹⁄₁₂
½ + ¾		

Sum	Equivalent fractions	Answer
$1/3 + 5/6$		
$4/5 - 1/2$		
$1/5 + 1/4$		
$1/4 - 1/5$		

COMMENTS ON THIS EXERCISE ARE GIVEN ON PAGE 107.

How do I multiply and divide fractions?

How often in your life have you had to multiply or divide two fractions outside a mathematics lesson? I suspect that the answer is, for most people, never. I therefore don't intend to devote much space to this difficult and rather pointless exercise. However, it *is* useful to know a few basic facts – for example, a half of a half is a quarter and a tenth of a tenth is one hundredth. There are also a few practical situations (like scaling the ingredients of a recipe, for example) where multiplication and division of very simple fractions is helpful. This is probably easier to understand by looking at decimal fractions, so we shall return to this topic in chapter 10.

Where fractions fit into maths

```
                              Whole numbers
                    Bits of  /
Maths—Numbers —— numbers ——— Fractions ——————— Equivalence
```

PRACTICE EXERCISES
(Try to spend up to thirty minutes on these exercises.)

FRACTIONS

1. Complete the blanks below.
   Share equally the following.

   Share 11 cakes amongst 4 people. Each gets → $2\frac{3}{4}$ cakes

   Share 17 cakes amongst 5 people. Each gets → ☐ cakes

   Share 5 cakes amongst 6 people. Each gets → ☐ cakes

   Share 20 cakes amongst 3 people. Each gets → ☐ cakes

2. There are 3 thirds in one. How many thirds are there in the following?

1	=	3	thirds.	10	=	___	thirds
4	=	___	thirds.	$3\frac{1}{3}$	=	___	thirds
5	=	___	thirds.	$7\frac{2}{3}$	=	___	thirds

3. Write the appropriate fractions on to the slices of the clock. Now add the fractions together. (Check that they add to 1.)

4. The fraction $\frac{8}{10}$ can be written more simply as $\frac{4}{5}$. Write the following fractions in their simplest form.

Fractions	$\frac{8}{10}$	$\frac{4}{6}$	$\frac{5}{10}$	$\frac{12}{18}$	$\frac{6}{9}$	$\frac{4}{16}$	$\frac{8}{48}$	$\frac{9}{18}$	$\frac{2}{22}$
Simple form	$\frac{4}{5}$								

5.
(i) Change all the following fractions to twelfths. (One has been done for you.)
(ii) Now rank them 1–6 in order of size (1 = largest, 6 = smallest). (One has been done for you.)

Fractions	$\frac{2}{3}$	$\frac{3}{4}$	$\frac{2}{6}$	$\frac{7}{12}$	$\frac{5}{6}$	$\frac{1}{2}$
Fractions as twelfths	$\frac{8}{12}$					
Rank	3					

ANSWERS ARE GIVEN ON PAGE 270.

Helping your child understand fractions

In this, the calculator age, common fractions are of much less importance than they used to be. Manipulating fractions is no longer a particularly useful skill – it is much easier to convert the fractions to decimals and use a calculator. However, simple fractions like halves, thirds, quarters and tenths are part of our everyday vocabulary and children should learn to use these fraction words with understanding. If calculators are being used, however, children need to know roughly what answer to expect. Thus they should know that $\frac{2}{3}$ is *about* 65 per cent (or 0.65) and that one third of fifty is *about* 16. They should be able to relate a common fraction to its corresponding decimal fraction and percentage (see next two chapters).

Fractions arise when things are being cut up and shared amongst a number of people. Sometimes you find yourself scaling up or down the ingredients of a recipe and again fractions are useful. If your child is around she can help you judge the appropriate angle at the centre of the apple pie or calculate how much flour is needed for six people given the appropriate amount for four (half as big again). Does she know what fraction of her family is adult, male, has blue eyes . . . ?

A game which my own children have enjoyed is to sing the song 'Ten Green Bottles' with the twist that fractions of a bottle are allowed to fall off the wall. Taking it in turns to sing the verses can be quite amusing, for if you are to keep up with the music you have about five seconds to perform the subtraction.

For example:

This chapter closes with a brief checklist of the sort of things your child should aim to understand about fractions.

FRACTIONS CHECKLIST
The ability to:

- know how many halves, thirds, quarters, etc. make a whole
- know that $13/3 = 4\frac{1}{3}$ and that $3\frac{3}{4} = 15/4$
- know that $2/3 = 4/6 = 6/9 = \ldots 20/30 = \ldots$
- select $\frac{1}{4}$ or $\frac{2}{3}$ of the sweets from a bag of sweets
- know that $\frac{1}{2}$ is bigger than $\frac{1}{3}$ but less than $\frac{3}{4}$
- know the value of $\frac{1}{2}$ of $\frac{1}{2}$ (answer $= \frac{1}{4}$)
- know how many $\frac{1}{3}$s in 2, $\frac{1}{4}$s in 5, etc.

- be able to perform + and − with simple fractions
- use fractions in a range of practical situations

ANSWERS TO EXERCISES FOR CHAPTER 9

Exercise A

(i)   4
(ii) 6
(iii) 4/6 or 2/3

Exercise B

(i)

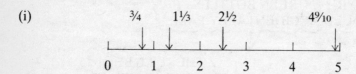

(ii)

Fraction	Equivalent fractions
3/4	6/8, 9/12, 12/16
6/18	1/3, 2/6, 3/9
2/5	4/10, 6/15, 8/20
10/20	1/2, 2/4, 3/6

Your answers here may be different from mine but still correct.

FRACTIONS

Exercise C   Adding and subtracting fractions

Sum	Equivalent fractions	Answer
$2/3 + 1/4$	$8/12 + 3/12$	$11/12$
$1/2 + 3/4$	$2/4 + 3/4$	$5/4$ or $1\tfrac{1}{4}$
$1/3 + 5/6$	$2/6 + 5/6$	$7/6$ or $1\tfrac{1}{6}$
$4/5 - 1/2$	$8/10 - 5/10$	$3/10$
$1/4 - 1/5$	$5/20 - 4/20$	$1/20$

10
Decimals

No doubt, like me, you find five fingers on each hand (well, four fingers and one thumb) a reasonable number to possess. Any fewer and we wouldn't be able to play 'Moonlight Sonata' with the same panache; any more and we'd have a bit of a struggle putting on a pair of gloves. It may not surprise you that the link between the number of our *ten*tacles and decimals is, well, more than *ten*uous. What I'm really saying, then, is that the reason our number system is based on the number ten is because humans have counted on their ten fingers for thousands of years.

'Decimals' (from the Latin *deci-* meaning ten) is really a way of describing the 'ten-ness' of our counting system. However, it usually refers to decimal fractions. And there is no shortage of those around us. Just listen to sports commentators, for example:

. . . the winning time of 10.84 seconds smashes the world record by two hundredths of a second.

. . . a long jump of 8 metres [point] 21.

. . . the winning scores for the pairs ice skating are as follows: 5.9, 5.8, 5.9 . . .

Decimal points appear whether we are talking about money (£8.14) or measurement (2.31 metres) and will appear on a calculator display at the touch of a button. Now I'd like you to find yourself a calculator and we'll discover more clearly what exactly a decimal fraction is.

What is a decimal fraction?

A decimal fraction is simply another way of writing a common fraction. What I'd like you to do is use your calculator to discover how fractions and decimals are connected.

For each box of questions below:

(a) write down the answers in fractions
(b) use your calculator to find the answers in decimals
(c) complete the blank in the 'Conclusion' box

Exercise A

Key sequence	(a) Fraction	(b) Decimal	(c) Conclusion
1 ÷ 2 =	½		The decimal for
2 ÷ 4 =	2/4		½ is 0.5
5 ÷ 10 =			
50 ÷ 100 =			
1 ÷ 4 =			The decimal for
2 ÷ 8 =			¼ is
25 ÷ 100 =			
3 ÷ 4 =			The decimal for
6 ÷ 8 =			¾ is
75 ÷ 100 =			
1 ÷ 10 =			The decimal for
10 ÷ 100 =			1/10 is

Now use your calculator to find the decimal values of the fractions in the table below.

Table 1

Fraction	½	¼	¾	1/10	1/5	2/5	3/10	9/10	1/20	1/8	1/3
Decimal											

COMMENTS ON THIS EXERCISE ARE GIVEN ON PAGE 116.

Let's now turn to the way we can represent decimal fractions on a number line. Since fractions and decimals are really very similar, it is not surprising that they can both be represented in the same way. For example, the fraction ¾ and the decimal 0.75 share the same point on the number line. Thus:

Having made a connection between fractions and decimals, the next exercise (called 'Guess and Press') gets you working just with decimals. The idea is to write down your *guess* as to what the answer will be for each sum. Then you *press* the sum on the calculator and see if you are right. The aim of this exercise is to help you see the connection between decimals and whole numbers.

DECIMALS

Exercise B

Sum	Guess	Press
0.5 ⊞ 0.5 ⊟	1	1
0.5 ⊠ 2 ⊟		
0.25 ⊠ 4 ⊟		
0.5 ⊠ 10 ⊟		
4 ⊟ 10 ⊟		
0.1 ⊞ 0.1 ⊟		
0.1 ⊞ 0.1 . . . ⊟		

10 times
1 ⊟ 10 ⊟

COMMENTS ON THIS EXERCISE ARE GIVEN ON PAGE 116.

What is the point of the decimal point?

If the world contained only whole numbers, we would never need a decimal point. The rules of place value tell us that the last digit of a number shows how many *units* it contains, the second last digit gives the number of *tens* and so on. For example, the number twenty-four is written as:

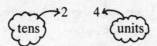

However, when we start to use numbers which include bits of a whole (i.e., with decimals) some other 'places' are needed – for the tenths, hundredths and so on. *The decimal point is simply a marker to show where the units (whole numbers) end and the tenths begin.*

You'll get a better idea of this by discovering when the decimal point appears on your calculator. As you do Exercise C, watch out for the decimal point . . .

Exercise C

Complete the blanks and then check with your calculator.

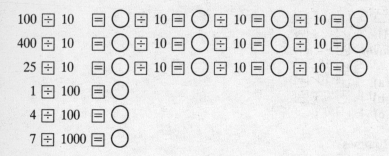

THERE ARE NO COMMENTS ON THIS EXERCISE.

You may have got a picture of the decimal point jumping one place to the left every time you divide by 10. But really the decimal point is nothing more than a mark separating the units from the tenths. Below I've written out the more complete set of 'place values' extending beyond hundreds, tens and units into decimals.

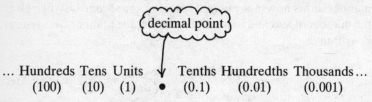

How do I use the four rules with decimals?

If you aren't sure, why don't you experiment with your calculator? You will quickly discover that the four rules work in exactly the same way for decimals as for whole numbers. However, there *are* one or two surprises.

Exercise D will get you started with addition and subtraction.

DECIMALS

In chapter 9 I suggested that you could learn about multiplying and dividing fractions yourself by experimenting with your calculator. Exercise D is designed to help you do just that.

Exercise D

The first column in this table gives you four multiplication sums involving fractions. For each sum:

(a) change the fractions to decimals (column 2)
(b) use your calculator to multiply the decimals (column 3)
(c) change the decimal answer back to a fraction (column 4)

		⟨Use your calculator⟩	⟨Use Table 1 if necessary⟩
(1) Sum in fractions	(2) Sum in decimals	(3) Decimal answer	(4) Fraction answer
½ × ½	0.5 × 0.5	0.25	¼
½ × ⅕			
⅗ × ½			

Now look at columns (1) and (4). When you think you have spotted the rule for multiplying fractions, write it below.

To multiply two fractions _____

_____

_____

COMMENTS ON THIS EXERCISE ARE GIVEN ON PAGE 116.

Let's now look at the following multiplication:

¾ × ⅖

As before, these two fractions can be converted into decimal form, thus:

0.75 × 0.4

giving an answer of 0.3, or ³⁄₁₀.

However, if we apply the rule given on page 116 and multiply the top and bottom numerals of each digit together, the answer is $\frac{3 \times 2}{4 \times 5}$ or ⁶⁄₂₀. There is no need to panic, however, because ⁶⁄₂₀ and ³⁄₁₀ are equivalent fractions and therefore have the same value.

Dividing fractions is a more painful and less useful skill so I don't propose to cover it here. This topic is one of several 'casualties' of the calculator age which has now nearly dropped off the school curriculum. And most of us would say 'Phew! Good riddance!' However, if you feel so moved you might like to set up a table like that in Exercise D and explore it for yourself.

PRACTICE EXERCISES

1. Mark the numbers 0.35 and 0.4 on the number line below.

   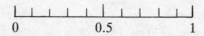

   Which of the two numbers is bigger_____?

2. In 0.6, the 6 stands for 6 _____ ?

3. Ring the number nearest in size to 0.78
   0.7 / 70 / 0.8 / 80 / .08 / 7

4. Multiply by 10:   5.49→ _____

DECIMALS

5. Add one tenth: 4.9 → _____

6.
   ```
      |  |  |  |  |  |  |  |  ↓ |
      21                      22
   ```

   This number is about _____ ?

7. How many different numbers can you write down between 0.26 and 0.27?

ANSWERS ARE GIVEN ON PAGE 271.

Where decimals fit into maths

```
                    Whole numbers
                   /                     Fractions
Maths — Numbers <      Bits of numbers <
                                         Decimals
```

ANSWERS TO EXERCISES FOR CHAPTER 10

Your calculator should have provided you with most of the answers to these exercises. However, here are some of the main points.

Exercise A

You should have been able to conclude that the decimal values of these fractions were as follows:

Fraction	1/2	1/4	3/4	1/10	1/5	2/5	3/10	9/10	1/20
Decimal	0.5	0.25	0.75	0.1	0.2	0.4	0.3	0.9	0.05

Fraction	1/8	1/3
Decimal	0.125	0.3333333

Exercise B

This exercise should help to give you a better feel for what a decimal is. The answers were

1; 1; 1; 5; 0.4; 0.2; 1; 0.1

Exercise C

No comment.

Exercise D

The rule for multiplying fractions is as follows:

To multiply two fractions, say 3/5 and 1/2, multiply the top numbers (3 × 1), then multiply the bottom numbers (5 × 2). The answer in this case is:

$$3/5 \times 1/2 = 3/10$$

DECIMALS

Helping your child to understand decimals

Eventually, after a great deal of experience your child should come to appreciate that decimal numbers are a natural extension of our whole number system. What this means in practice is being able to understand place value. So just as you add up to ten units and then swap them for one ten, so you add up to ten hundredths and swap them for one tenth. This is illustrated in the two addition sums below:

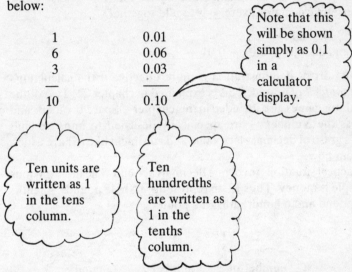

```
  1        0.01
  6        0.06
  3        0.03
 ──        ────
 10        0.10
```

Note that this will be shown simply as 0.1 in a calculator display.

Ten units are written as 1 in the tens column.

Ten hundredths are written as 1 in the tenths column.

She should (eventually!) realize that 'the more digits a number has, the bigger it is' was valid when she was using whole numbers but is not true for decimal numbers. For example, many, if not most, eleven-year-olds believe that 24.91257 is a bigger number than 83. They are easily impressed by a long string of digits and tend to ignore completely the position of the decimal point. It might help if we wrote whole numbers with the decimal point included after the units digit (for example, 83. or 83.0). One way of emphasizing place value to your child would be to write the number 83 as 8^3 or the number 24.9 as $24_{.9}$.

Children need to see beyond a string of digits and get a sense of how big the number actually is. For example, it is probably more useful to know that 24.91257 is between 24 and 25 (or just less than 25) than to quote it to five decimal places. Sensibly used, calculators are an excellent medium for getting this message across. For example, ask your child to put a large number into the calculator display and then repeatedly divide by 10. At first she can watch what happens. Later she can try to predict what will happen. Then try multiplying a small decimal fraction by 10, 100, 1000 and so on.

A useful calculator exercise is to add together

0.1 + 0.1 + 0.1 + . . . =
Then add 0.01 + 0.01 + 0.01 + . . . =

This sort of demonstration is more effective if the calculator's *constant facility* is used. This is explained in chapter 22. Two of the calculator games also included in this chapter – 'Space Invaders' and 'Guess the Number' – are specifically designed to improve children's grasp of decimal place value and you might like to have a look at them now.

Practical situations involving decimals abound. The most obvious example is money. Thus £3.46 represents 3 whole pounds, 4 tenths of a pound and 6 hundredths of a pound.

However, the money representation of decimals can be confusing. We *say* £3.46 as 'three pounds forty-six', rather than 'three point four six', which is the more correct decimal form. This latter version emphasizes the decimal place value of each digit. Otherwise children get into trouble when dealing with sums of money like one pound and nine pence which they inevitably write as £1.9, rather than £1.09.

DECIMALS

Children grow up with decimals and metric units all around them. However, we still have feet and inches, pounds and ounces, and these are the units that most adults feel happy with. When we eventually move completely to metric units (and it must happen soon, surely!) both you and your child will begin to use them naturally when cooking, doing carpentry, measuring up curtains or laying a carpet. Then I feel sure that much of the current fear of decimals will simply vanish.

Finally, here is a checklist of the sort of things your child should aim to know about decimals.

DECIMALS CHECKLIST

The ability to:
- know that the 4 in the number 6.143 refers to four hundredths
- mark decimal numbers on the number line
- arrange decimal numbers in order from smallest to biggest
- multiply and divide decimal numbers by 10, 100 and 1000
- know that 3.45 is half way between 3.4 and 3.5
- know that 0.25 means ¼ and 3.75 means 3¾
- handle units (metres, kilograms, pounds [£] in practical situations)
- know roughly what answer to expect in a calculation involving decimals

11
Percentages

It is clear that most people find percentages difficult. A recent government report on maths (reference 3, page 277) has confirmed that amongst adults there was '. . . a widespread inability to understand percentages'.

Yet this is despite the fact that you can't pick up a newspaper or watch TV without hearing the word percentage over and over. Here are just three examples from one page of my daily paper.

- Unemployment amongst school leavers is now running at nearly 50 per cent.
- In the year ending July, cheese consumption rose by 4 per cent in volume and 15 per cent in value. Over 96 per cent of British families now regularly buy cheese.
- The average foreign visitor spends 50 per cent more in Britain than the average Briton spends abroad.

Most people are able to make some sense out of these sort of statements. For example, they can see that youth unemployment is high and that the vast majority of British families put cheese on their weekly shopping list. But not everyone will have a clear understanding of exactly what is meant by the third extract. (Put another way it means that for every £100 the average Briton spends abroad, the average foreign visitor spends £150 here. What we mean by 'average' here is also not clear, but that's another story. . . .)

The reason for the confusion is, I think, quite simple. Many adults and most children don't really understand what a percentage is.

What is a percentage?
The first thing you should realize about a percentage is that it is very

PERCENTAGES

similar to a decimal and a fraction. Like them, a percentage is used to describe a 'bit' of a number. But really it is nothing more than a particular sort of fraction.

Think back to chapter 9 on fractions. I talked about how fractions could be *equivalent*. Here are four fractions which are equivalent:

$$\tfrac{1}{2},\ \tfrac{2}{4},\ \tfrac{5}{10},\ \tfrac{50}{100}$$

These fractions are equivalent because they share the same value – of a half.

Now look at the last of these four fractions: $\tfrac{50}{100}$. You might read it as 'fifty out of a hundred'. A shorthand way of saying this uses the Latin words *per centum* meaning 'out of every hundred'. So $\tfrac{50}{100}$ is the same thing as 'fifty per cent'.

(out of)   (a hundred)

Well, if a half (i.e., $\tfrac{50}{100}$) is the same as 50 per cent, what do you think a quarter becomes as a percentage?

The answer is 25 because $\tfrac{1}{4} = \tfrac{25}{100}$, or 25 per cent.

How do I change a fraction to a percentage?

By now you might have worked this out for yourself. If not, here is the method which I've summarized in a flow chart for convenience – and to impress on you how easy it is!

A flow chart showing how to change a fraction into a percentage

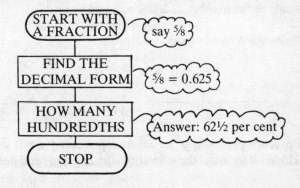

Exercise A   *Changing fractions to percentages*

Fill in the blanks in the table below.

Fraction	Decimal fraction	Percentage
½	0.5	50
¾		
7/10		
⅕		
1/20		
⅗		

ANSWERS TO THIS EXERCISE ARE GIVEN ON PAGE 130.

The quick way of writing 'per cent' is %. So 15% is really another way of writing 15/100 or 15 per cent.

Like fractions and decimals, percentages can be represented on the number line. In the percentage number line below, 100% corresponds to the number 1, 200% to 2 and so on.

Percentage Number Line

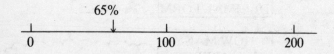

In practice percentages are rarely represented in this way but I've included it to stress the similarity with fractions and decimals.

PERCENTAGES

Why do we bother with percentages?

The main advantage of percentages is that they are much easier to compare than fractions. For example, which do you think is bigger, ¾ or ⁷⁄₁₀? Written like this you can't really say, because the slices of the whole 'cake' (quarters and tenths) are not the same size. To make a proper comparison, the fractions need to be broken down to the same size of slice, and hundredths are very convenient. So here goes. . . .

Fraction	Percentage
¾	75%
⁷⁄₁₀	70%

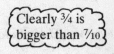

Clearly ¾ is bigger than ⁷⁄₁₀

If we look at a practical example you will get a better idea of how useful percentages are.

Which of the following would represent the bigger price rise?

(a) Milk to go up by 2p per pint
(b) A refrigerator to go up by £5

In one sense the answer could be (b), because £5 is more than 2p. But since you and I buy many more pints of milk that we do fridges, we would probably be more concerned if milk went up by 2p per pint. Try converting these price rises to percentages and a very different picture emerges. Do Exercise B now.

Exercise B

Complete the table below.
(I've taken the price of milk to be 20p per pint and the refrigerator to cost £100.)

	Cost	Price rise	Relative increase as a percentage
Milk	20p	2p	
Fridge	£100	£5	

ANSWERS TO THIS EXERCISE

I hope you agreed with my calculation that milk went up in price by 10 per cent whereas the fridge went up by only 5 per cent in price.

How would I calculate percentage reductions and increases?

Unless the percentages convert to *very* simple fractions (e.g., 100 per cent or 50 per cent, 25 per cent or 10 per cent), I would always use a calculator. Here are two examples.

Example (a)

Socks – old price 80p a pair – now 50% off!

This one is easy! Since 50% is ½, there is a reduction of half of 80p. This is 40p.
So the new price is 80p − 40p = 40p

Example (b)

Estimate for kitchen alterations – £226 + 15% VAT

I asked three friends to work out the *total* cost (including VAT) of my kitchen alteration. They all did it by a different method! Here are two of them:

Method (i)   Kath works out VAT a lot in her job. Without a calculator, she prefers to find 10% and 5% separately – then add them together.

PERCENTAGES 125

Kath's method

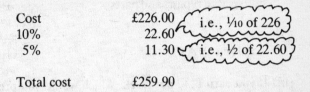

Cost	£226.00
10%	22.60
5%	11.30
Total cost	£259.90

Method (ii) Peter isn't too confident with percentages. However, he can do it with the help of a calculator. He works out the VAT first and then adds on the 226.

Peter's method

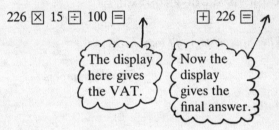

Neither Peter nor Kath's solution is the most efficient possible, but what mattered was that they had found a method which they understood and could use.

By the way, you might like to think what the most efficient calculator method *is* to work out the total cost of my kitchen alteration. My own method is given on page 126. Look at it before you go on to Exercise C.

Exercise C   Finding percentage increases and decreases

(i) A table normally sells at £42. How much will it cost with a 30% reduction?

(ii) Check my garage bill

R. Nixon Motors

Full service repairs and parts	£86.20
VAT (15%)	16.93
Total (inc. VAT)	£103.13

(iii) If you earn £130 per week, which would you prefer?
   A rise of (a) 6%
       or (b) £6 per week?

(iv) Your taxable earnings are £584 this month. How much will you have left after paying 30% to the taxman?

COMMENTS ARE GIVEN ON PAGE 130.

Where percentages fit into maths

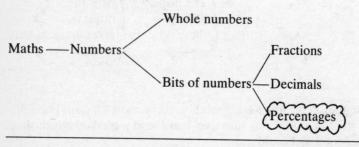

My method for adding on 15% VAT is to multiply the given amount (226) by 1.15.* So to calculate the total cost (inc. VAT) of my kitchen alterations I will use this key sequence on the calculator:

226 ☒ 1.15 ☐

*This comes from $\frac{10 + 15}{100}$ which equals 1.15. It may not be obvious to you where the 1.15 comes from. It is helpful here to think in terms of hundredths. Before VAT we have $^{100}/_{100}$ of the given amount. Adding 15% will increase this to $^{115}/_{100}$, which equals 1.15.

PERCENTAGES

PRACTICE EXERCISES

1. Which is bigger, 8% or ⅛? _____

2. Which is bigger, 15% or 1/15? _____

3. If the rate of inflation drops from 5% to 4%, are prices:
   (a) going up
   (b) coming down
   (c) neither? _____

4. What is 20% of £80? _____

5. What is 10% of 20% of £80? _____

6.
	Old price (per doz.)	New price (per doz.)
Size 5 eggs	81	85
Size 2 eggs	94	98

   Which has had the greater price increase, size 5 or size 2 eggs?

   _____

7. My garage bill has come to £20.93 and includes VAT at 15%. What would the bill be:

   (a) without VAT? _____

   (b) with VAT at 25% instead of 15%? _____

8. Why do children's sweets tend to suffer greater inflation than most of the other things we buy?

   _____

   _____

ANSWERS ARE GIVEN ON PAGE 272.

Helping your child to understand percentages

It will be no surprise to most of you that a lot of children's time in school takes place with one eye shut and the other staring out of the window. Many children come away from a lesson in percentages (or whatever) with only a few pieces of the jigsaw and have to somehow try to fill in the rest of the picture themselves. Unfortunately they don't always get it right . . .

Can you spot where this child has gone wrong?

The problem is that she has started from a true fact that 10% = $1/10$ and built up a rule which doesn't work for any other fraction.

This is probably the most common misapprehension about percentages and one you might like to talk to your child about. You may find that if she has problems it all can be traced back to a fuzziness about fractions. Your child may know that 20% is more than 5%. However, she needs to be sure that $1/20$ is less than $1/5$ to realize that these fractions do *not* correspond, respectively, to 20% and 5%. Shopping together during the sales is an opportunity for your child to discover that $1/3$ off is a better discount than 10% off. She will also learn that $1/3$ off the price of something fairly cheap like a packet of envelopes represents only a small saving in actual money. Are the family earnings keeping pace with inflation? Is her pocket money keeping pace with inflation? Why is it that the price of cheap children's goods go up by a greater percentage than other goods costing more?

For children who *have* a basic understanding of what percentages are, many have difficulties in calculating with them in a practical example. There is no one correct method for calculating with percentages. Children, like adults, will develop confidence in their own method, given time and opportunities to cope with percentages in a variety of everyday situations. However, even more important than being able to calculate with percentages is knowing when and why we use percentages at all. The following example illustrates the point:

> There are about 23 million telephones in USSR and only 16 million in Canada. Therefore people in Russia are better off in respect of telephones than the Canadians.

The facts are right but the conclusion is wrong when you realize that the population of the USSR is about eleven times that of Canada. In fact about 65 per cent of Canadians have a telephone, whereas only 9 per cent of Russians have one.

Perhaps the most important thing your child needs to grasp is that fractions, decimals and percentages are really the same thing. I have found that drawing three number lines side by side is helpful for children to spot these connections. Thus:

Fraction	Decimal	Percentage
0	0	0
½	0.5	50
⅘	0.8	80
1	1	100

The dotted line shows that ⅘, 0.8 and 80% have the same value.

PERCENTAGE CHECKLIST

The ability to:

- realize that percentages make it easier to compare fractions
- link percentages to common and decimal fractions
- convert from percentages to (simple) fractions and decimals and vice versa, e.g., 75% = ¾ = 0.75
- express something as a per cent of something else, e.g., 6 is 25% of 24
- calculate percentage increases and decreases

ANSWERS TO EXERCISES FOR CHAPTER 11

Exercise A

Fraction	½	¾	7/10	⅕	1/20	⅗
Decimal fraction	0.5	0.75	0.7	0.2	0.05	0.6
Percentage	50	75	70	20	5	60

Exercise B

No comment.

Exercise C

(i) The reduction in price is 30% or 3/10.
3/10 of £42 is £12.60, so the reduced price is £42 − £12.60 or £29.40.
(Note: a quicker way of doing this is to say that a 30% reduction will bring the price down to 70% of the old price. 70% of £42 = £29.40.)

(ii) The VAT is incorrect.
Using a calculator:

86.20 ⊠ 0.15 ⊟ (12.93)

So I've been overcharged by £4.
(Note: a direct method for checking the final bill is to press:
86.20 ⊠ 1.15 ⊟.)

(iii) 6% of £130 = £7.70, which is a bigger rise than £6.

(iv) I will have 70% of £584 left, which is £408.80.

12
Basic arithmetic

Well, I hope that you aren't feeling too shell-shocked after six chapters of heavy-duty basic arithmetic! The aim of this chapter is to let you take stock of the arithmetic which you have read and, I hope, tried out over the past six chapters. The first exercise below is a fairly traditional test of the skills already covered. Please do the exercise and then mark your work using the answers on page 135. After you have marked your own efforts, Exercise B will give you the chance to play 'teacher' by marking a child's attempt at the same exercise. The child, whom I have called Anne, doesn't actually exist but her answers reveal the common mistakes that children make. (I am grateful to the work of the CSMS [Concepts in Secondary Mathematics and Science] research project, from which several of the questions and errors given below were taken or adapted. See reference 4, page 277.)

Exercise A

1. (a) Add
   29
   16
   46
   —

   (b) Add
   381 + 41 + 104

   (c) Subtract
   50
   21
   —

   (d) Subtract
   twenty-two from forty-eight

2. (a) Multiply
    18
   × 5
   ——

   (b) What are 4 lots of 68?

   (c) Divide
    6|228

   (d) What is 6 divided by 12?

3. (a) How many apples do I add to 4 apples to get 10?

   (b) A box of sweets is divided among 21 people so that each person gets 3 sweets. How many sweets were in the box?

   (c) Make up a short word problem like (a) and (b) using the numbers 8 and 4 and a *division* sum.

4. Tick which of the following is true or false (*don't* work out the sums in each case).

	True	False
(a) 469 + 287 = 287 + 469		
(b) 469 − 287 = 287 − 469		
(c) 469 × 287 = 287 × 469		
(d) 469 ÷ 287 = 287 ÷ 469		

5. (a) Shade in three quarters of the circle.

(b) Add
$3/4 + 2/5$

6. (a) My petrol tank holds 9½ gallons.
How many gallons does it hold when half full?

(b) A fridge costing £60 is sold at ⅓ off.
How much is the sale price?

7. (a) Write down which you think is the biggest and the smallest of these three numbers.
Say why you chose them.

$$0.65 \quad 0.60 \quad 0.7$$

_____ is the biggest because _____
_____ is the smallest because _____

(b) Multiply 2.46 by 10.

(c) In a gymnastics competition, Vera scored:
9.7, 9.9, 9.6 and 9.4
What was her average score?

8. (a) A coat costs £40. In a sale it is reduced by 5%. How much does it now cost?

(b) In a test, 203 out of 765 fourteen-year-old children were able to answer question 8(a) correctly. What is this as a percentage?

BASIC ARITHMETIC

ANSWERS TO EXERCISE A

1. (a) 91   (b) 526   (c) 29   (d) 26

2. (a) 90   (b) 272   (c) 38   (d) ½

3. (a) 6   (b) 63
   (c) '8 sweets are shared among 4 people. How many did each person get?'

4. (a) T   (b) F   (c) T   (d) F

5. (a)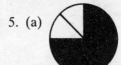

   (b) ¾ + ⅖ = ¹⁵⁄₂₀ + ⁸⁄₂₀ = ²³⁄₂₀ or 1³⁄₂₀

6. (a) 9½ ÷ 2 = 4¾ gallons
   (b) sale price = 60 × ⅔ = £40

7. (a) 0.7 is the biggest because it has the most tenths.
       0.60 is the smallest because it has the least tenths and no hundredths.
   (b) 2.46 × 10 = 24.6
   (c) Vera's average score = $\frac{9.7 + 9.9 + 9.6 + 9.4}{4}$ = ³⁸·⁶⁄₄ = 9.65

8. (a) Reduction = 40 × ⁵⁄₁₀₀ = £2.50 therefore sale price = £38
   (b) $\frac{203}{765}$ × 100 = 26.5%

Exercise B

As you will see below, Anne's attempt at Exercise A isn't very successful. In fact, she gets most of her answers wrong. Look carefully at her working and see if you can discover why she is having problems. The chances are that your child will have similar difficulties in many cases.

Homework   Anne 11

1(a) 29   (b) 381   (c)   50   (d)   408
      16   41   − 21   − 202
    + 46   + 104   31   206
    ─────   ──────   ────   ─────
      82   895

2(a)   18   (b)   68   3 r 48   d Ans 2
       ×5   68 68   (c) ──────
     ────   ×× 68   6) 228
      540   24 +68
                      ──────
                       272

3(a) 10 add 4   (b)   7   (c) I had 8 marbles and
      10   ──────   my brother had 4. So we
    + 4   3) 21   divided them. Then we had 6 each
     ────
      14   Ans 7 sweets

4 all four are true

5(a)   5(b) $\frac{3}{4} + \frac{3}{5} = \frac{5}{9}$

6(a) Half of 9 = 4½   (b) $\frac{1}{3}$ of £60 = £20
     Half of ½ = ¼   Ans = £40
     Ans 4¾

7(a) 0.65 is the biggest number because 65 is bigger
than 60 or 7. 0.7 is the smallest number because 7 is
smaller than 65 or 60

  (b)   2.46   (c) 9.7 + 9.9 + 9.1   8a $\frac{1}{5}$ of 40 = 8
        10   Ans about 9.7   Ans £32
       ─────
        0.00
        2.46
       ─────
        2.46
                    (b) ?

BASIC ARITHMETIC 137

Finally, you might like to compare your detective work with mine. Anne has left a few clues on her answer sheet which reveal quite a lot about her difficulties with maths. The errors in the first four questions are particularly common among seven- to eleven-year-olds. The harder questions at the end reveal the mistakes of children of eleven years and over.

1. (a) Anne has correctly added her units column to 21. However, instead of writing 1 and carrying the 2 to the tens column, she did the reverse. Why, I wonder? Perhaps she has mislearnt (and therefore not understood) the procedure here and thinks she should write down the larger digit and carry the *smaller* one to the tens column. (This rule actually works most of the time.)
   (b) Anne hasn't arranged the digits into the right columns. I'll make a mental note to check whether this was just through sloppy presentation or because she is unsure about place value.
   (c) If only subtraction were so easy! Anne has made up a simple rule for subtraction which is this:

   > For each column, subtract the smaller digit from the larger.

   I need to remind her what subtraction really is – with counters if necessary.
   (d) This is a common mistake amongst children of about eight years old. She will eventually see that we don't write the number 'twenty-two' as we say it. Anne needs lots of practice at seeing and saying numbers; a calculator will help greatly.

2. (a) This is really the same mistake as 1(d), the basic problem being place value. I'll get her to play a game of calculator Space Invaders and then encourage *her* to chat about the hundreds, tens and units columns.
   (b) At least Anne knows that the sum to do is one of multiplication but that it can also be done, albeit rather

clumsily, by an addition sum. I think Anne needs some practice with the 4 times table.

(c) Trouble with place value again!

(d) Anne is so used to textbook examples 'working out' exactly that she doesn't read the question very carefully. This is more common in problem-type questions like 3(b). Also she may be mixing up 'divided by' with 'divided into'.

3. (a) Possibly all Anne saw of this sentence was the following:

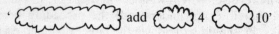

So she did!

(b) As with 2(d), Anne looked at the numbers, rather than trying to make sense of the story line of the question. The word 'divided' combined with the numbers 21 and 3 were all she chose to read.

(c) To Anne, dividing means splitting equally. It is one of a number of words which children use in a different sense to its mathematical use.

4. No comment.

5. (a) Anne got this one right.

(b) This mistake of adding the tops and the bottoms of fractions is very common. It shows basic misunderstanding of fractions and I've booked in a lesson on equivalent fractions for Anne tomorrow.

6. (a) and (b) Well done, Anne! She does have good common-sense grasp of fractions when they appear in an *everyday* problem.

7. (a) This is another common misunderstanding. Anne must be discouraged from reading decimals like 0.65 as 'nought point *sixty-five*'. I will find her a calculator activity based on decimal place value.

(b) It is important to understand what happens when you multiply or divide numbers by 10, 100, 1000, etc. Anne clearly needs some help with this.

(c) Although stuck with the maths, 9.7 was a good common-sense estimation. I must encourage this in her.

8. (a) Anne seems to think that 5% means ⅕. Oh dear, back to the drawing board.

(b) The large numbers made this a particularly difficult one and I'm not surprised she gave up.

Most of the errors which Anne made seem to suggest that, like many children, she has learned certain procedures by rote, but has little understanding of the concepts of numbers, decimals, fractions and so on. The solution is *not* to tell her that she is stupid or lazy or to give her lots of practice at what she can't do. Not only will this not work but it will also create the sort of anxiety and fear of maths that many of us remember from our own school days. There is really no alternative but to go back to the basics. Give her a variety of physical representations (slices of 'cakes' for fractions, for example). Get her *doing* and *talking* before she goes back to writing things down on paper. Provide various practical contexts (half an apple is easier to understand than the fraction ½). Where practice is needed, give her a game or an exploration on the calculator.

It should also be stressed that the questions in the test paper were extremely artificial. Many of them were specifically designed just to trip children up. For their own part children soon discover that there are ways of beating the system and avoiding having to think too hard. For a start, they can look at the title of the exercise. If it says 'multiplication of decimals' then you can safely ignore the words in the problem and simply multiply the two numbers contained in the sentence. In cases where the title of the exercise is less of a giveaway, children learn to scan the text for 'cue words' such as 'shared' or 'times' and then perform these operations on the numbers in the question. A third trick is to look at the numbers themselves for a clue as to what calculation is expected. If the two numbers are, say, 33 and 11, then it is clear that, whatever the words say, the answer to the question must be 3 (i.e., 33 ÷ 11).

Most children, then, don't struggle to make human sense of arithmetic problems. Given the artificiality of many of these questions, however, one can't really blame them! Here are two examples:

- Peter shares out his 24 sweets among three friends . . .
  (a likely story!)

- Look at this list of birthdays. Write down the difference between the oldest person and the youngest person.
  (Who would ever want to know this?)

Such questions are a test of *how* to calculate rather than a test of a pupil's understanding of *which* particular calculation to choose. Real life problems don't come prepackaged in this way and real life answers tend not to work out exactly. The next chapter tackles this broader question of 'Knowing what sum to do'.

13
Knowing what sum to do

Previous chapters have concentrated on numbers and number skills. But all the number skills in the world are useless if your child doesn't know *when* to use them. In this chapter we will be less concerned with the arithmetic of the sum than knowing what sum to do.

Imagine that your child has been given £1 to buy sparklers for Bonfire Night. Bought singly they cost 12p each. Your child wants to know how many she could expect to get for her £1, but doesn't know how to work it out. Think about the problem yourself and make a note of what sum you would do to answer it.

There is no one single answer to this question. In fact the problem can be solved in (at least) four different ways; one for each of the four rules. Which one of the following did you choose?

(a) Adding: How many 12 pence can I add together to get to £1?

(b) Subtracting: How many 12 pence can I take away from £1?

(c) Multiplying: What must I multiply 12p by to get to £1?

(d) Dividing: What is £1 divided by 12p?

As long as your choice of method gives the right answer, it can be used, but one of these four methods is the most direct and most efficient – the last one, (d). (However, it is better for children to use an inefficient method which they understand than an efficient one which they don't.)

Sometimes it happens that you and your child search through the mathematical skills at your disposal and *can't* come up with one which helps solve the problem. Then you are STUCK!

What can I do when I'm stuck?

If the problem really does involve maths that you can't use, then I'm afraid you just have to ask someone for help. However, this is rarely the case. The vast majority of everyday problems can be solved by simple maths. Also, as you have already seen with the sparklers example, if you can't do a problem by one method, you can probably find another which will work.

A common experience is to feel 'bogged down' in a maths problem; stuck, but not sure why or where. A simple remedy for this is to *break the problem up into bits*. Most problems tend to fall into these three stages:

(a) *Deciding* what sum to do
(b) *Doing* the sum
(c) *Interpreting* the answer

Exercise A asks you to use these three stages in the sparklers problem given at the beginning of the chapter.

Exercise A   The stages of problem-solving

(a) Work out on a calculator or on paper how many 12p sparklers you would get for £1.

(b) Now think about the three stages and jot down below what you did at each stage.

Stage 1 DECIDING_____

_____

2 DOING_____

_____

3 INTERPRETING_____

_____

KNOWING WHAT SUM TO DO

COMMENTS ON THE EXERCISE.

Here are my answers for Exercise A.

DECIDING –   I want to do a division sum. But dividing 12p into £1 gets me nowhere. I must change the £1 into 100 pence so that the units are the same.

DOING –   The sum on the calculator is:

100 ÷ 12 = 8.3333333
                    answer

INTERPRETING – I will get 8 sparklers. The bit after the decimal point tells me that I can expect some change – but I'm not sure how much yet.

By breaking problems up into bits like this you will get a better idea of why you are stuck. Is it because you are using a wrong method (deciding), or because the arithmetic is hard (doing), or perhaps that you can't make sense of your answer (interpreting)?

This chapter concentrates on just the first of these stages. Here now are four tips which will help you through the 'deciding what sum to do' stage.

Tip 1 Ask yourself 'What information do I need?'

This week I bought a tin of chocolate powder at the supermarket. There was a choice of two makes. Since I assumed that they were of the same quality, I was interested in choosing the tin which gave the better value for money. Jot down in Exercise B what information I need, to be able to decide.

Exercise B   *What information do I need?*

_____

_____

_____

COMMENTS ON THE EXERCISE

Since I am interested in value for money, I need to know the *price* and the net *weight* of each tin.

Finding out what information you need is an important first step in most practical problems.

Tip 2 Boost your confidence with a calculator

The figures on price and weight of the two tins were as follows:

	Price	Weight
Tin A	81p	500 g
Tin B	60p	420 g

Having got the information, you may still be at a loss as to what sum to do. This could be because the numbers look hard and you are starting to panic about the arithmetic that lies ahead. There is no doubt that it is easier to see how to do a problem when the numbers are simple. One way of boosting your confidence at this point is to pick up a calculator. Knowing that the machine will take care of the tedious 'doing' arithmetic, you can concentrate all your energies on deciding what buttons to press – i.e., what sum to do. You can try out different sums on your calculator. Fortunately, it won't complain about being sent on a fool's errand!

> **Tip 3** Make up a simpler (similar) problem

If you still don't know which buttons to press on your calculator, then there is clearly an important missing link. The chances are that you haven't got a clear picture of the story line of the problem. The way to tackle this is to *make up a simpler but similar problem*. For example:

Is it better to buy three pencils for 9p or four pencils for 10p?

This simplified version of the problem does the same sort of job as a parable in the Bible or one of Aesop's fables. When the details of a story are stripped to the bare bones, you can see the underlying principle more clearly. What 'moral' can you draw from the simplified 'pencil' problem? Well, answer the question and see!

Clearly it is better to buy four for 10p because they work out at only 2½p each compared with 3p. Ah! The 'moral' of this simple problem is that you compare value for money by dividing price by quantity. This gives what is known as *unit price* (i.e., price per unit). If you look at the price labels on the foods in supermarkets you will sometimes see the unit price stated as well as the price and quantity. For example, the packet of bacon in my fridge gives this information on the label:

Weight 9 oz   Price 74p   Price per lb £1.35

And now let's put the theory into practice by doing the original problem.

Exercise C

Use your calculator to calculate the unit prices of the two tins of chocolate powder. Then decide which is better value for money.

Complete the following:

	Price (p)	Quantity (g)	Unit price (pence per gram)
Tin A	81	510	
Tin B	60	420	

Tin ☐ *is better value for money because it has a bigger/smaller unit price.*

COMMENTS ON THE EXERCISE

According to my calculator, Tin A came out with a unit price of 0.159 pence per gram, whereas B proved better value at only 0.143 pence per gram. You will notice that I rounded my answers here to 3 places after the decimal point. (Rounding will be taken up again in chapter 14.)

No doubt some of you were rather unhappy with those long decimal strings on the calculator. If so, you may have preferred to tackle this problem upside down! What I mean by this is that instead of dividing price by quantity for each tin you could have divided quantity by price – like this:

	Number of grams per pence
Tin A	510 ÷ 81 = 6.296
Tin B	420 ÷ 60 = 7

Doing the problem this way round, we choose the tin which has the *bigger* number of grams per pence. Not surprisingly, Tin B works out better value whichever way round we do the calculation.

> **Tip 4** Ask yourself, 'Will this sum give a silly answer?'

THIS SUM GIVES A SILLY ANSWER

Finally, once you've decided on a sum, what about that answer? Does it feel right? Would you expect to order 300 metres of curtain material or pay 26p for your electricity bill? Often you don't have to complete the sum before realizing that it is going to give a silly answer. This notion of relating the maths to common sense is an obvious but important help in deciding what sum to do.

But now, what about children? How confident is your child at using maths to solve practical problems?

Is it an add, Mum?

Most children are not good at knowing what sum to do. One reason is that they find it difficult to know *how* they have solved a simple problem, even when they got it right. Take this example:

12 sweets can be arranged with 3 sweets per row.
How many rows are there?

Most children will realize that the answer is 4. But why 4? Their answer to *this* question will either be:

'It's obviously 4' or perhaps
'Because three fours are 12'

This second response is interesting when you compare it with the 'correct' adult response, which is:

'Because 12 divided by 3 equals 4.'

The original problem was actually one of division, but the most common explanation given by twelve-year-olds is that of multiplication. Earlier in the chapter I talked about how the method for solving an easy question can help you tackle a harder one. Children often need help to recognize what method they have used, even when they have answered an easy question correctly.

A second reason why children are not good at recognizing what sum to do is very simple – they don't get enough practice at doing *real* sums. Textbook 'problems' come prepackaged. Exercises tend to provide questions which test one skill only (written at the top of the page). The questions themselves provide neither too much nor too little information. The answers which the children produce in class rarely have to be matched up against common sense ('Is my answer silly?').

The remedy is to encourage children, both at home and at school, to tackle the realistic practical problems that surround them. It is only when children make real decisions and choices that they begin to put school-learnt skills to proper use.

Like adults, children also need support and encouragement if they are to learn to use maths as naturally as they read and write. Although confidence grows with practice, the four tips listed in the previous section should be as helpful for your child as they are for you. For with maths, as with most problems in life, the best starting point is learning to ask the right questions.

Summary

This lesson has moved away from the *learning* of number skills and concentrated instead on learning how to *use* them in practical problems.

The following four tips were suggested for when you are stuck:

- Ask yourself, 'What information do I need?'
- Boost your confidence with a calculator.
- Make up a simpler (similar) problem.
- Ask yourself, 'Will this sum give a silly answer?'

14
Measuring

It is said of frogs that they classify any other animals that they meet into just three categories.

If it is small, they eat it.
If it is large, they run away from it.
And if it is about their own size, they mate with it.

I think it's fair to say that, in most cases, humans are more discriminating! However, although frogs may not be engaged in highly sophisticated measuring here, they *are* trying to understand the bigness or smallness of things around them. Any activity which does this, however imprecisely, can be called measuring.

Measuring is rightly seen as an important topic in school maths. Thinking back to the three main branches of maths mentioned in chapter 6, measuring has strong links with two of them – numbers and space. The network diagram below shows how measuring fits into the overall picture.

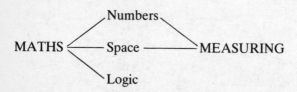

What do we measure?

Most people tend to think of measuring as using weighing scales or a tape measure. But what sort of thing do these devices tell us about?

MEASURING 151

The *dimensions* being measured here are, respectively, weight and length. The *units* used may be pounds (or kilograms) for weight, and inches (or centimetres) for length. There are, of course, many dimensions other than length and weight which we need to measure – for example:

   temperature, time, area, capacity, angle, volume, speed . . .

Exercise A will give you a chance to think about those dimensions and also the units in which they are usually measured.

Exercise A

Complete the table (the first two have been done for you).

Question	Dimension	Units
How heavy is your laundry?	weight	kg or lb
How long is the curtain rail?	length	cm or in
How hot is the oven?		
How far is it to London?		
How fast can you run?		
How long does it take to cook?		
How much does the jug hold?		
How big is your kitchen?		
How big is the field?		

COMMENTS ON THIS EXERCISE ARE GIVEN ON PAGE 159.

As you see, I have included both the imperial and the metric units, since both are still in common use. Because many people are confused by these various units of measure, I have explained them in chapter 15.

Why do we measure?

Quite simply we live in a more complex world than a frog. Although words like 'large' and 'small' are sometimes good enough ('Give me some of the *large* apples', 'I'd like a *small* helping'), often we need to be more precise. Here is an example where the word 'large' proved inadequate during the national rail stoppage in July 1982.

> A very large proportion of the men didn't show up for work. (ASLEF spokesman)
>
> A large proportion of the men showed up for work. (British Railways spokesman)

The question here is, how large is large?

You may have seen signs on the motorway advising drivers of 'large' vehicles to stop at the next emergency phone and contact the police. Rest assured that the small print below goes on to explain that 'Large means 11'.00" (3.3 m) wide or over'. The reason that we tend to measure with *numbers* is to help us make decisions and comparisons fairly and accurately. Careful measuring helps us bake 'the perfect cake' every time, lay well-fitting carpets with a minimum of waste, check that the children's shoes don't pinch and so on. Exercise B asks you to think about the measuring dimensions involved in these sorts of everyday tasks.

Exercise B

Tick which dimensions are being measured in the following. (There may be several dimensions involved in each activity.)

Activity	Length	Area	Volume	Weight	Time	Temperature	Capacity	Speed
1 Baking a cake								
2 Buying and laying a carpet								
3 Checking the children's shoes								
4 Setting out on a journey in good time								

COMMENTS ON THIS EXERCISE ARE GIVEN ON PAGE 159.

How do we measure?

Descriptions can come in two basic forms. Descriptions of *quality* tend to be made with words, while those of *quantity* involve numbers. Measurement is usually associated with the second of these, numbers, but not always. A number of descriptive words like large/small, fast/slow, chilly/warm, are ways of indicating whereabouts on the scale (of size, speed and temperature) they lie. Yet, though they make no mention of numbers these descriptions are a sort of measure. Words which describe a colour, on the other hand, do not normally relate to a useful scale – you can't say that red is more than blue, or give a sensible answer to the question 'how green is my valley?' – and so are simply descriptions. Words which can be *ranked* into a meaningful order produce what is called an *ordering scale*. Exercise C will give you practice at using an ordering scale.

Exercise C

Rank the following words into a useful ordering scale and say what dimension is being measured by this scale. (The first one has been done for you.)

Words	Rank order (low → high)	Dimension
(a) 1. likely 2. impossible 3. doubtful 4. certain 5. highly improbable	2. 5. 3. 1. 4.	likelihood
(b) 1. jog 2. stop 3. sprint 4. walk		
(c) 1. walk 2. lie 3. sit up 4. stand 5. roll over		
(d) 1. wool 2. linen 3. silk 4. cotton 5. rayon		

COMMENTS ON THIS EXERCISE ARE GIVEN ON PAGE 160.

MEASURING

To summarize, then, measuring can take the following three basic forms:

- words alone
- words which can be ranked in order
- numbers

The types of measuring scale which these three approaches use are:

- words (not really a scale)
- ordering scale
- number scale

Although all three types of scale are helpful in providing an interesting variety of descriptions and comparisons, measuring with numbers is the most important.

How accurately should we measure?

The accuracy with which we measure depends entirely on what and why we are measuring. A brain surgeon and a tree surgeon have different needs for accuracy when sawing up their respective 'patients'. A nurse weighing out drugs will exercise greater care and precision than a greengrocer weighing out potatoes. Calculators may give eight-figure accuracy but often the numbers on which the calculation is being performed are only approximate. For example, suppose you wish to know how many panels of fencing (1.8 m each in length) to buy for your garden which is 21 m long. The calculator will produce an answer 11.666667. In this sort of case it is plainly silly to give an answer to eight figures. If the last six digits of your answer are either dubious or unnecessary, then dispose of them. However, you have to be a bit careful how you do this. Numbers can be shortened so that you finish with a suitable number of digits (say three). This is called giving your answer 'correct to three significant figures'. With the above example, calculating the number of panels of fencing requires an answer correct to two significant figures – in this case twelve panels. (Note, however, that you would still need twelve panels even if the answer on the calculator was 11.33333!)

This process of simplifying unnecessarily accurate measurements to a near approximation is called *rounding*. Some examples are given in Table 1 below.

Table 1

Measurement	Rounded measurement correct to 3 significant figures (3 sig. figs.)
4.1834926	4.18
371.41429	371
0.0142419	0.0142
74312.692	74300
81.869143	81.9

The last example in the above table is different to the others in the following respect. As you can see in column 2, the third digit has been *rounded up* from an eight to a nine. The clue to why it has been rounded up can be found by looking at the fourth digit in the original number; the six. Since it is bigger than five, the eight in the tenths column is *rounded up* to a nine.

Exercise 1 (page 157) at the end gives you practice at rounding. (It is easier to do than to read about!)

There are many practical situations where careful measurement is quite inappropriate; where an *estimate* is good enough. For example, when extending my kitchen recently I read that planning permission tends to be automatically granted provided the volume of the extension does not exceed $70m^3$. I estimated that the dimensions of the extension were no more than $5 m \times 3 m \times 2.5 m$. Clearly this contained a volume of much less than $70 m^3$ so I could proceed with the plans. Estimation is a skill which greatly improves with practice. I sometimes find it helpful to imagine everyday objects of a standard size to help me make an estimate. For example:

- height or distance – how many 6′ policemen laid end to end?
- capacity/volume – how many pints of milk?
- weight – how many 1 kg bags of sugar?

PRACTICE EXERCISES

Exercise 1   Rounding

Round the following numbers to four significant figures.

	Number	4 sig. figs
(1)	4124.7841	4125
(2)	38.4163	
(3)	291.7412	
(4)	39042.611	
(5)	39048.619	
(6)	38.4131	
(7)	446.982	
(8)	0.142937	
(9)	1317.699	
(10)	3050.1491	

Exercise 2

The calculator work for each of the four problems below is the following:

$$30 \boxed{\div}\ 9\ \boxed{=}$$

giving an answer of 3.3333333.

However, the actual answer to each problem is different. See if you can match each problem to its appropriate answer.

Problems	Answers
1. A cake serves 9 portions. How many cakes must you buy for 30 people to get one portion each?	
2. If you wish to share 30 sweets amongst 9 children, how many sweets will each child get?	3.33 3⅓ 3 4
3. A boxed set of 9 records costs £30. How much does that work out at per record?	
4. One pint of wine exactly fills 9 glasses. How much wine would fill 30 glasses?	

ANSWERS ARE GIVEN ON PAGE 273.

MEASURING 159

ANSWERS TO EXERCISES FOR CHAPTER 14

Exercise A

Question	Dimension	Units*
How heavy is your laundry?	weight	kg or lb
How long is the curtain rail?	length	cm or in
How hot is the oven?	temperature	degrees (°C or °F)
How far is it to London?	length (or distance)	km or mile
How fast can you run?	speed	km/hour or mph
How long does it take to cook?	time	minute
How much does the jug hold?	capacity	cc or pint (or fl oz)
How big is your kitchen?	volume	m^3 or ft^3
How big is the field?	area	hectare or acre

*I will explain most of these units in chapter 15.

Exercise B

Activity	Length	Area	Volume	Weight	Time	Temperature	Capacity	Speed
1 Baking a cake				√	√	√	√	√
2 Buying and laying a carpet	√	√						

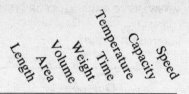

	Length	Area	Volume	Weight	Time	Temperature	Capacity	Speed
3 Checking the children's shoes	√	√						
4 Setting out on a journey in good time	√				√			√

Exercise C

	Rank order (low → high)	Dimension
(a)	(2) impossible (5) highly improbable (3) doubtful (1) likely (4) certain	likelihood
•(b)	(2) stop (4) walk (1) jog (3) sprint	speed
(c)	(2) lie (5) roll over (3) sit up (4) stand (1) walk	stages of child development
(d)	(5) rayon (3) silk (1) wool (4) cotton (2) linen	temperature (on an electric iron!)

Summary

Measuring is an important topic in mathematics and a vital life-skill.

This lesson focused on the 'what', 'why' and 'how' questions of measurement. These can be summarized as follows:

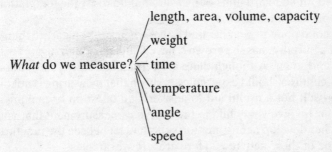

Why do we measure? To help make decisions and comparisons.

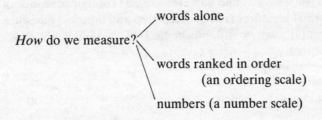

Helping your child to measure

There are many aspects of measuring that children find difficult or confusing – so many that much of the next chapter will be devoted to looking at some of them. However, there are two key features of measuring that you should bear in mind when engaging in measuring activities with your child. The first is that the standard units of measurement can seem very complicated, particularly since we currently use two systems of units – metric and imperial. Although your child will eventually need to know how many millimetres there are in a centimetre and so on, at the age of six or seven years she will probably be encouraged in school to devise her own units of measure. So she may find it easier (and more fun) to measure her height in matchboxes or discover that her bag of crisps weighs the same as fourteen little plastic cubes. These experiences are important as they help your child to discover dimensions, like weight, length, area, capacity, for herself. Eventually she will see the need for standard units of measure but do resist the temptation to impose these too soon.

A second point to bear in mind is that when adults measure things it is for a purpose: has she grown?, have I added enough sugar?, will it fit?, and so on. Although there are certain measuring skills to be learnt, children should experience early on that measuring is not an end in itself but a useful aid to answering a question or making a decision. Involve your child in the everyday measurement that you do, let her develop her estimation skills and let her see the practical outcome of these activities. Here are a few examples:

- cooking and baking. These activities cover most of the measures your child will ever use and contain a balance of formal measures (ml, g, and so on) and informal measures (cupful, teaspoonful, pinch, etc.).
- carpentry and general DIY
- dressmaking and curtain-making
- laying tiles, carpets, etc.
- making scale models

When the occasion arises, encourage your child to think about and talk about the questions raised at the beginning of the chapter. For example:

- how big is 'big'?
- what do the words we use to describe 'size' actually refer to?
- how accurately do we need to measure?

and so on.

To finish this chapter you might be amused at the following classroom anecdote which took place when a friend was conducting a lesson on measuring.

> My third-year maths set aren't known for their flashes of insight and brilliance. In fact they have yet to register on the Richter

MEASURING

scale of . . . well, practically anything. A recent topic under their scrutiny was 'accuracy' and I was working hard trying to give it a little 'pzaz'. This involved finding examples drawn from life where the term 'significant figures' had some significance.

'Take the population of Greater London,' I began. 'You can't really say that it has a population of *exactly* 6,877,143 because people are moving in and out of it every day.'

After some debate, 3c seemed to accept that perhaps 'just under seven million' was a more useful and more realistic figure for London's population. Now, I thought, let's see if the idea has really been understood. I decided to test them with a similar problem which was closer to home.

'How many people live in Milton Keynes?'

To my great surprise and delight, about ten hands shot up. I wasn't used to this sort of enthusiasm in response to a question. I nodded to the normally silent lad in the corner.

'Well Andrew?' I invited.

'I do, Miss,' replied Andrew.

15
More measuring

Chapter 14 introduced various common dimensions of measurement – length, weight, time and so on. I tried there to explain when and how we use them. This chapter looks at some of those dimensions more closely, particularly length, area and volume, and the sort of tangles that children can get into when they use them. First, however, a word or two about the units that we use in measuring.

Imperial and metric units

Until about 1970, measurement in this country was largely done with *imperial* units (from the good old days of the Empire). In recent years, however, we British have at last owned up to the fact that we have the same number of fingers and thumbs as the rest of the world, and have 'gone decimal'. The *decimalization* of money in 1971 was carried out quickly and effectively. As a result most people mastered the new coinage within days. It was also intended that we abandon the familiar imperial units of length, weight and capacity within a few years. This change, called *metrication*, was to have swept away the most familiar of our measuring units – feet, inches, yards, pounds, stones, pints, gallons and so on – in favour of metres, kilograms, litres and the like. Indeed, during the 1970s, many children learnt only the metric units in school on the assumption that the old imperial units would soon be six feet (sorry, 1.83 metres) under. Unfortunately, however, the changeover was so half-hearted that, at the present time of writing, we are still regularly using both systems (and having fun trying to convert from one to the other!). Teachers are now being asked to teach children both systems in school.

MORE MEASURING

Conversions between metric and imperial units tend to involve rather awkward numbers. For example, there are 'about' 39.370078 inches in one metre! Not surprisingly, a number of half-baked approximations have appeared like the metric yard (39 inches), the metric foot and even the metric brick.

Have a look at the table below and you will see just how 'approximate' some of the approximations are.

Unit	True value
Metric yard = 39 ins	39.370078 ins
Metric inch = 2 cm	2.54 cm
Metric foot = 30 cm	30.48 cm
Metric mile = 1500 m	1609.344 m

In this chapter I'll explain the various units, both metric and imperial, as we look at each dimension of measurement in turn, starting with the most basic measure of all – length.

LENGTH

The measurement of length is so commonplace that we have quite a choice of words available to us to describe it. For example, measuring upwards we call *height*. *Depth* refers to measurement downwards. And if we want to measure sideways, we have *width* or *breadth*. The list could be added to with *distance* for large lengths and, perhaps, *gap* or *interval* for tiny ones. This choice is all part of the richness of adult language but, as you might imagine, for children trying to master the basics of measurement it can be most confusing.

One of the first measuring skills a child will learn is to use a ruler or tape measure. Here children often make the mistake of measuring the object in question from the 1 cm mark, rather than the 0 cm mark on their ruler.

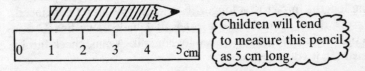

166   HELP YOUR CHILD WITH MATHS

If your child consistently measures lengths which are one inch (or one cm) too short, you can suspect that this is the reason.

Another common difficulty that children have when measuring length is shown up in Exercise A below. Many adults also have problems of this type, so try the exercise yourself before reading on.

Exercise A

Compare lines B and C with A and say whether you think they are longer, shorter or the same length.

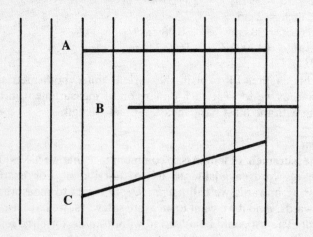

B is longer than/shorter than/the same length as A
C is longer than/shorter than/the same length as A

COMMENTS ON THE EXERCISE
Most children in the eleven- to fourteen-year age range will realize that although line B sticks out further than line A, it is actually shorter. However, up to half of the children of this age think that because A and C stretch between the same 'tramlines', they are the same length. In fact C is longer than A. If you aren't convinced, measure them yourself with a ruler. If I had drawn line C at a steeper angle it would have been longer still. Again, check this by measuring if you aren't sure.

MORE MEASURING

UNITS OF LENGTH

On page 173 of this chapter you will find a table summarizing all the metric and imperial units that you and your child are likely to need. I'll explain how to use it by looking at just this extract below, which deals with units of length.

Table 1

Imperial units

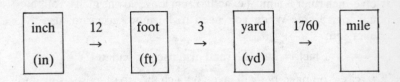

Metric units

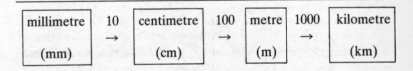

The numbers on the arrows tell you how to convert from one unit to another. Thus, there are 10 mm in 1 cm, 100 cm in 1 m, and so on. If you want to know how many mm are in 1 m, then *multiply* the two numbers 10 and 100 (i.e., there are 1000 mm in 1 m).

It will help you understand and remember the metric units when you realize that:

- for each dimension there is a *basic unit* – the basic unit for length is the metre
- all the other units get their name from the basic unit: e.g., because *centi-* means one hundredth ($1/100$), then a *centimetre* is one hundredth of a metre

Table 2 will help you work out the others.

Table 2

    MILLI- → one thousandth (¹/1000)

    CENTI- → one hundredth (¹/100)

    DECI- → one tenth (¹/10)

    KILO- → one thousand (1000)

Converting *between* metric and imperial units is a little trickier. If you don't need to be too accurate, it is helpful to remember that a twelve-inch ruler is almost exactly 30 cm long. So one inch is roughly equal to 2.5 cm. When you need to be more accurate, use the conversion:

    1 inch = 2.54 cm (and also use a calculator!)

Exercise 1 on page 174 will give you a chance to practise using and converting these units of length.

AREA

Length is a *one-dimensional* measurement because it normally involves only one direction. Problems involving surfaces (size of paper, carpet, curtain material . . .) are *two-dimensional*. Sometimes we describe area simply by stating the length and the breadth. For example, curtain material is bought by the metre (length) but we also need to know that the roll is 1 m 20 cm wide.

If you are buying paint, on the other hand, the instructions on the tin may say something like: '. . . contents sufficient to cover 35m^2 . . .' A 'm^2' or a 'square metre' is the basic metric unit of area. It means exactly what it says. One m^2 is the area of a square, 1 m by 1 m.

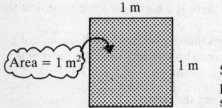

So there should be enough paint to cover 35 of these squares.

Note that the particular shape I've drawn here is a metre square. A square metre can be any shape.

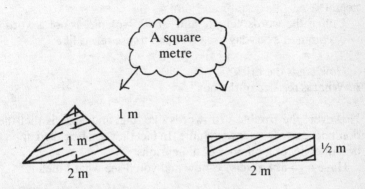

Similarly, a ft² (square foot) is the area of a square 1 ft by 1 ft, and so on for the various other units of area. However, it's all not quite as easy as it sounds. Have a go at Exercise B now and see if you can avoid the traps that children often fall into.

Exercise B

(i) How many square feet (ft²) are there in one square yard (yd²)?

(ii) How many cm² are there in one m²?

(iii) What is the area of this rectangle?

(iv) If you double the dimensions of the rectangle above (i.e., the length *and* the breadth) what do you do to the area?

COMMENTS ON THIS EXERCISE ARE GIVEN ON PAGE 177.

VOLUME

If you ask most children about the word 'volume', they will tell you that it is the knob on the TV set which makes it go loud and quiet.

'Volume', as used in maths, is rather different. It describes an amount of space in *three dimensions* (3D). If you think of a box, its volume will depend on the three dimensions length, breadth and height.

Unlike the words 'length' and 'area', 'volume' is not a word in very common everyday use. We tend to use terms like:

How big is the brick?
or What is *the size* of the box?

However, the trouble with words like 'big' and 'size' is that they don't necessarily refer to volume. In fact they can be called upon to describe any of a number of dimensions.

Have a go at Exercise C now and you'll see what I mean.

Exercise C

Jot down the dimension that you think the word 'size' refers to in each of the following:

	Dimension
(i) The size of a pencil	
(ii) The size of a piece of paper	
(iii) The size of a packet of washing powder	
(iv) The size of a bucket	

COMMENTS ON THIS EXERCISE

I felt that in none of those cases was volume the appropriate dimension. Probably they were the following:

(i) Length
(ii) Area (either given by a standard size like A4 or by stating the dimensions)
(iii) Weight
(iv) Capacity

People who work in the building trade become skilled at estimating amounts of earth and concrete. They usually measure these volumes in so many 'cubes'. A 'cube' usually refers to a *ft³ or a cubic foot* although more recently it tends to refer to a cubic metre. A cubic foot is the amount of space taken up by a cube measuring 1 ft by 1 ft by 1 ft:

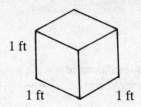

Similarly a cubic centimetre (cc) is the amount of space taken up by a 1 cm cube.

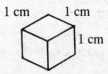

The difficulties that children have with volume are similar to those with area, only more so. Exercise D will give you a flavour of some of them.

Exercise D

(i) Which of the two shapes, A or B, has the bigger volume?

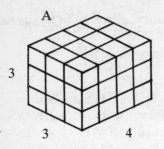
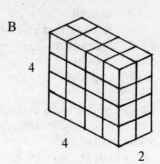

(ii) Why is this advert misleading?

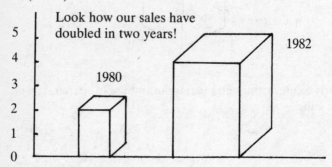

COMMENTS ON THIS EXERCISE ARE GIVEN ON PAGE 178.

Like volume, capacity is also a measure of space in three dimensions. The difference is that capacity describes a container and refers to how much the vessel holds. The units of capacity, which can be found in Table 3, are normally only used with liquids.

Table 3 Metric and imperial units

	Imperial Units	Metric Units
Length	inch (in) —12→ foot (ft) —3→ yard (yd) —1760→ mile	millimetre (mm) —10→ centimetre (cm) —100→ metre (m) —1000→ kilometre (km)
Area	in^2 —144→ ft^2 —9→ yd^2 —4840→ acre —640→ square mile	cm^2 —10,000→ m^2
Volume	in^3 —1728→ ft^3 —27→ yd^3	cm^3 —1,000,000→ m^3
Capacity	fluid oz —5→ gill —4→ pint —2→ quart —4→ gallon	ml —1000→ litre (with cm^3 —1000→ litre)
Weight	oz —16→ lb —14→ stone —112→ cwt —20→ ton	gram (g) —1000→ kilogram (kg) —1000→ tonne
Speed	miles per hour (mph)	kilometres per hour (kmph)

This means
10 mm = 1 cm
100 cm = 1 m
1000 m = 1 km

These are the BASIC units

Table 4 Converting between metric and imperial units

	Accurate conversion	Rough 'n' ready conversion
Length	1 in = 2.54 cm 1 m = 39.37 cm	1 inch = 2½ cm 1 metre is just over a yard – a very long stride
Capacity	1 gallon = 4.54 litres 1 litre = 1.76 pints	1 gallon = 4½ litres 1 litre = 1¾ pints – a large bottle of orange squash
Weight	1 lb = 0.454 kg (or 454 g) 1 kg = 2.2 lbs	1 lb = just under ½ kilo 1 kg = just over 2 lbs – a bag of sugar
Speed	100 mph = 161 kmph 100 kmph = 62.1 mph	5 mph = 8 kmph

PRACTICE EXERCISE

1. Use Tables 3 and 4 (and a calculator where appropriate) to answer the following:

(i) Which of these would be a reasonable weight for an adult?
60 kg, 600 kg, 6 kg

(ii) Is it cheaper to buy a 25 kg bag of potatoes or a 56 lb bag for the same money?

(iii) What is the height of your door in metres?

(iv) Some French roads have 90 kmph speed limits. What is this roughly in mph?

(v) Is a ½ litre of beer more or less than a pint?

(vi) If we bought milk by the ½ litre, how many bottles would you have to buy to have roughly 4 pints?

(vii) A recipe asks for 100 g of butter. Roughly what fraction of a ½ lb packet will you cut off?

ANSWERS ARE GIVEN ON PAGE 273.

Helping your child to measure

Given at the end of the last chapter were a number of practical activities which use measuring – baking, DIY, and so on. I know from my own experience that involving young children in these household tasks can sometimes be frustrating; but if I am prepared for slow progress it can also be fun. I always try and explain to my children why and what we are measuring at each stage. You could ask your child if she can find you the implements needed to measure the various commodities you are dealing with (volume of milk, length of wood, and so on). If you can bear it, let her pour the milk into the measuring jug and hold the end of the tape measure, etc. If you are well organized in advance and have cleared sufficient working space for you both, there shouldn't be too many disasters!

Don't worry about exposing your child to the formal units of measure (cm, g, and so on) before she learns about them in school. These are part of living and even at the age of three or four years she will benefit from seeing you use them purposefully in a practical situation. At around the age of six or seven years children can get the measuring bug. Your child might enjoy recording her personal body measurements in a book. Remind her of it six months later and she will be intrigued to see by how much she has grown.

By the time they are nine or ten children might enjoy making a scale drawing of their house. This will involve measuring each room, choosing a suitable scale (say 1 m = 2 cm) and then making separate scale drawings of the ground floor and the first floor. I have always done this for a new house which I am about to move into. My children have helped me make 2D cardboard representations (to scale) of the important pieces of furniture – bed, piano, chest of drawers, sofa, and so on. We've then had a lot of fun 'trying out' items of furniture in various rooms and positions. It is actually quite useful to help you confirm that the chest of drawers doesn't quite fit into the alcove on the top floor *before* you drag it up the stairs!

Try to take every opportunity to develop your child's estimation skills. Before she measures the length of the wall or the height of the window or the weight of a partly used bag of sugar, ask her first to estimate what answer she expects. She will probably be 'miles' off initially but as she gets a better sense of the important units of measure her estimation skills will improve. It is only when she knows roughly what answer to expect that she will notice mistakes like measuring a plank of wood as 87 cm rather than 2 m 87 cm.

Estimating can be made into a game. For example, you give your child a length or a weight to estimate – then measure it and see how close she is. You can take turns at doing this and see if you both improve. Give your child practice at using all the measuring instruments in the home – the bathroom scales, kitchen scales, measuring jugs, tape measures, and so on. Encourage her to think about and talk about the more difficult composite units like miles per hour and lbs per square inch.

Finally, here is a checklist of the basic skills of measuring which your child will need.

MEASURING CHECKLIST
The ability to:

- be familiar with the common measures
  (length, weight,* area, volume, capacity, time, speed, temperature)
- use and understand standard units of these measures
  (centimetre, kilogram . . .)

MORE MEASURING 177

- estimate lengths, weights and so on, in terms of these units – to know when a measurement is about right and what sort of tolerance is appropriate
- use various measuring instruments (tape measure, ruler, weighing scales, balance, measuring jug, thermometer, clock)
- be aware of composite units (miles per hour, price per gram, kg per cc, and so on)

ANSWERS TO EXERCISES FOR CHAPTER 15

Exercise A

No comment.

Exercise B

Answer	Comments
(i) $9 \text{ ft}^2 = 1 \text{ yd}^2$ (ii) $10{,}000 \text{ cm}^2 = 1 \text{ m}^2$	Children often give the answers 3 and 100. It would help them to draw a square yard and then count out the 9 ft^2 inside it.
(iii) Area = $6 \times 4 = 24 \text{ m}^2$	It's not immediately obvious to children that the area of a rectangle can be found by multiplying the length and the breadth.

*Note that, strictly speaking, we should talk about 'mass' rather than weight. The weight of an object is a measure of the force of gravity acting on it and this will vary depending on where the object is in relation to the earth. Mass is the 'amount of matter' which the object contains and, wherever its position, this will not vary. This distinction isn't usually introduced to children until they are about twelve or thirteen years, however, so don't let it worry you!

Answer	Comments
(iv) The area increases by four times.	Children usually think that the area must double also.

Exercise C

No comment.

Exercise D

Answer	Comments
(i) Volume of A = 3 × 3 × 4 = 36 units Volume of B = 2 × 4 × 4 = 32 units So A has the bigger volume.	If children don't know to multiply the dimensions, they usually count little cubes. This is decidedly tricky since some are invisible.
(ii) Although sales have only doubled, the 3D drawing suggests a volume which has grown by eight times.	Children and adults are often misled by this sort of sneaky advertising.

16
Shapes and solids

Although this book is concerned mostly with number, it shouldn't be forgotten that mathematics provides us with ways of understanding other important aspects of our life such as making sense of space. The next three chapters concentrate on some basic spatial ideas. This chapter looks at the bread-and-butter questions of school geometry – triangles, circles, angles and solids. Chapter 17 explores the idea of proportion while chapter 18 takes a mathematical look at some of the beautiful patterns in nature and art.

What are the common shapes?

The commonest simple shapes, like the triangle and square, are bounded by a certain number of straight sides. If the sides are all the same length the shape is said to be *regular*. For example, a five-sided shape is called a *pentagon*. A regular pentagon will look like this:

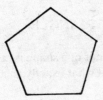

An irregular pentagon will look something like:

this   or this

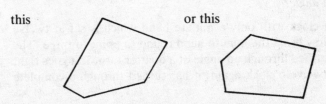

Table 1 below gives the name of each shape corresponding to the number of sides it contains.

Table 1

Number of sides	Name
3	Triangle
4	Quadrilateral
5	Pentagon
6	Hexagon

I have already distinguished between regular and irregular shapes. For triangles and quadrilaterals, however, the regular forms are sufficiently important to merit special names. These are called, respectively, the equilateral triangle and the square.

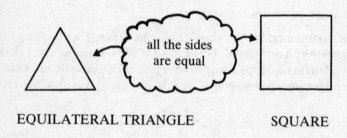

EQUILATERAL TRIANGLE   SQUARE

The sharpness of the corner of a shape is called an angle. However, we need to be clear about what exactly an angle is. The next section will explain this briefly.

What is an angle?

Imagine a clock with only a minute hand which is set at twelve o'clock. Now bring the minute hand round to point to three. The hand has turned through an angle of a quarter turn. If it goes right round to twelve o'clock again, it has turned through a complete turn.

SHAPES AND SOLIDS

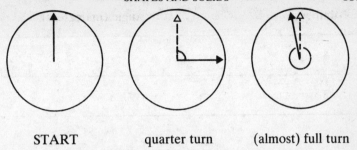

START   quarter turn   (almost) full turn

Angles *could* be measured in fractions of a turn but this is not a very sensible measure, especially for small angles. There are several acceptable ways of measuring angles, the most useful being *degrees*. One complete turn when the minute hand goes right round the clock face corresponds to 360 degrees. So a quarter turn = 90 degrees (usually written as 90°). Another name for 90° is a *right angle*.

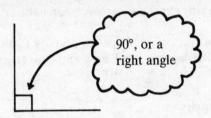

90°, or a right angle

Exercise A

If the minute hand of a clock starts at twelve o'clock, what angle (in degrees) has it gone through if it is turned in a clockwise direction to point to the following numbers?

Number	Angle (in degrees)
6	180
9	

Number	Angle (in degrees)
4	
3	
7	
11	
all the way round to 12 again	

COMMENTS ON THIS EXERCISE ARE GIVEN ON PAGE 193.

The three most important shapes in primary maths are triangles, circles and quadrilaterals. I will look at the first two of these here.

Exploring triangles

An important property of all triangles is that their angles always add to 180°. This is one fact about geometry that most of us are stuck with for life! Have a go at Exercise B now.

Exercise B

Cut out *any* shaped triangle from a piece of paper and mark the three angles, thus:

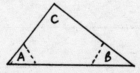

Now snip off two of the angles and place them against the third one. Check that those three angles together form a straight line (180°).

THERE ARE NO COMMENTS ON THIS EXERCISE.

The shape of a triangle depends on the size of its angles. You may have already noticed that equilateral triangles have equal angles as well as equal sides. Since these add to 180°, each of the angles of an equilateral triangle must be 60°. Another common triangle is the right-angled triangle, which, as the name suggests, contains one angle of 90°.

Perhaps the most interesting triangle is the right-angled triangle. As long ago as 2000 BC the Egyptians and Babylonians were using this triangle to solve problems. Their interest was probably based on the construction of buildings and the need to align bricks exactly at right angles. You might like to think about this problem yourself for a few minutes: how would *you* construct an angle of 90° without the aid of geometric instruments or a T square?

The method of four thousand years ago, using a knotted rope where the distances are marked out in the ratio 3:4:5, is sometimes still used today.

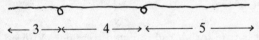

When this rope is stretched and pulled into a triangle thus,

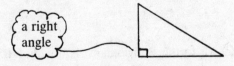

the angle opposite the longest side is a right angle.

In fact the 3, 4, 5 triangle is not the only one which produces a right angle. There are many more; for example:

5, 12, 13
8, 15, 17
7, 24, 25
. . .

As you might expect, these are not arbitrary sets of numbers. There is a mathematical connection between them. Have a go at Exercise C and try to discover the connection for yourself.

Exercise C

See if you can find a relationship which links this set of numbers:

3, 4, 5

(Hint: try squaring each of the numbers and see if that helps.)

COMMENTS ON THIS ACTIVITY

You may have spotted that when you add the squares of the two smaller numbers you get the square of the largest number. In other words, the square of 3 (9) plus the square of 4 (16) is equal to the square of 5 (25).

i.e., $3^2 + 4^2 = 5^2$

This result is true for *all* right-angled triangles. You might like to check for yourself that the following are true:

$5^2 + 12^2 = 12^2$
and $8^2 + 15^2 = 17^2$

This result, although known to the Babylonians, is usually attri-

SHAPES AND SOLIDS

buted to the Greek mathematician, Pythagoras (around 500 BC). When Pythagoras finally proved this theorem it is said that he was so excited that he sacrificed two thousand oxen to the god Apollo. If you managed to work out Exercise C without reading on to the solution, I suggest a packet of smoky bacon crisps all round to celebrate! Incidentally, the posh name for the longest side of a right-angled triangle is the *hypoteneuse*. A textbook statement of Pythagoras' theorem, then, would look something like this:

Pythagoras' theorem

In a right-angled triangle, the square of the hypoteneuse is equal to the sum of the squares of the other two sides.

Exercise D gives you some practice at using Pythagoras' theorem.

Exercise D

(Use your calculator.)
(i) Calculate the diagonal.

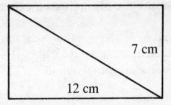

(ii) How big is the side of a square if its diagonal is 7.07 cm?

COMMENTS ON THIS EXERCISE ARE GIVEN ON PAGE 194.

Circles

You need to be told two facts about a rectangle – its length and its breadth – before you can draw it. What is special about a circle is that only one fact – its radius or its diameter – is enough to tell you all you need to know about it. Incidentally, the *radius* of a circle is the distance from its centre to the edge (the *circumference*). The *diameter* is the distance right across the centre of the circle and is therefore twice the radius.

What, then, is the connection between the diameter (or radius) and the circumference of a circle? If you draw a few circles of different size, measure the diameters (with a ruler) and circumferences (approximately with a piece of string) you may well conclude that the circumference is about three times the diameter. And you would be about right. This 'fact' has been known for thousands of years. Indeed there are several references in the Bible which suggest that early craftsmen expected this to be so.

> And he made a molten sea, ten cubits from the one brim to the other: it was round all about . . . and a line of thirty cubits did compass it round about.
>
> 1 Kings 7:23

In fact the ratio is slightly bigger than three. In Exercise E you can work it out a bit more accurately.

Exercise E

The table below gives the diameter and circumference of four circles measured fairly accurately (to the nearest mm). Find the ratio of circumference/diameter in each case and then find the average ratio of the four circles.

	Diameter (cm)	Circumference (cm)	Circumference/Diameter
Circle 1	10.0	31.5	
Circle 2	8.0	25.0	
Circle 3	7.0	21.4	
Circle 4	5.0	15.8	

Average ratio circumference/diameter = _____

COMMENTS ON THIS EXERCISE ARE GIVEN ON PAGE 194.

As you may have already realized, what you have been calculating is the value of the famous number pi. It is written as π (the Greek letter pi), and its true value can never be exactly calculated. This is because the decimal string goes on and on and on . . .

Here is its value to the first eight figures:

3.1415927 . . .

If you divide 22 by 7 on your calculator you get

3.1428571

which is pretty close to the true value of π. So the number $3\frac{1}{7}$ (i.e., $^{22}/_{7}$) is commonly used as a good approximation to π. An even closer approximation is $^{355}/_{113}$.

When it comes to finding the *area* of circles, the number π crops up again. You might like to explore this for yourself by drawing different-sized circles on to graph paper and estimating their area by counting up the little squares inside each circle. If you do this you will get something like the following:

	Radius (cm)	Area (cm²)
Circle 1	5	78.5
Circle 2	4	50.3
Circle 3	3.5	38.5
Circle 4	2.5	19.6

(Area cm² — This is the number of little cm squares inside the circle.)

For each of these circles, and indeed for any circle, the area can be found by squaring the radius and multiplying the result by π.

For example the area of Circle 1 is $5 \times 5 \times \pi$, which is about 78.5. Expressed as a formula the result can be written as:

$$A = \pi \times r^2$$

(A — the area of a circle; r — the radius)

Solids

The shapes that we have looked at in this chapter — circles, triangles and so on — are all two-dimensional. That means that they exist on a flat surface but have no depth. The real world has three physical dimensions and therefore the objects we tend to meet are solid. Many of the everyday objects which we deal with are fairly complex and difficult to analyse mathematically – for example, motorcars, chairs, guitars, and so on. However, there are a few basic solids which are worth looking at.

The commonest solids are:

- sphere (or ball)
- cube
- brick (or cuboid)
- cylinder

and
- cone

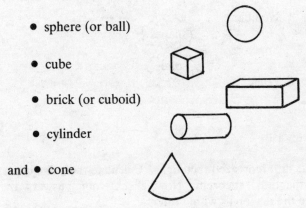

There is much to discover about these (and other) shapes.

One way of learning about their properties is to try to make them from cardboard. As you might imagine, this is a popular activity in the primary school – glue and paper cuttings everywhere! The systematic approach is to plan out what shape you want to cut out and draw it carefully on the cardboard. Such a drawing is called a *net*. Here is a net for a cube.

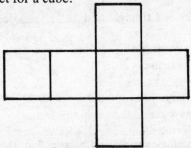

Try to imagine how this would fold up into the shape of a cube.

Drawing a net for a sphere is decidedly tricky! (Impossible in fact.) However, you might like to try to make a cube, a cuboid (brick) and a pyramid. (It is helpful if you add little tabs so that there is something to glue on to.)

Where shapes and solids fit into maths

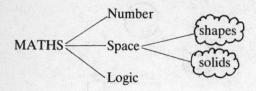

PRACTICE EXERCISES

1. Draw four different-sized rectangles. Calculate their diagonals using Pythagoras' theorem. Now check your answers by measuring the diagonals with a ruler.

2. Fill in the blanks in the following.

    (i) A right angle of _____ degrees.

    (ii) _____ right angles are the same as one complete turn.

    (iii) The angles of a triangle add to _____ degrees.

    (iv) The three angles of an equilateral triangle are all _____ degrees.

    (v) Another name for a regular quadrilateral whose angles are all equal is a _____.

    (vi) When the clock shows half past twelve, the angle between the minute hand and the hour hand is _____ degrees (careful! – this is trickier than it looks).

3. My bicycle wheels are 70 cm in diameter. How many turns will they make on a journey of 1 km?

4. A certain bus company insists that passengers may not carry with them parcels longer than four feet. You wish to bring a snooker cue on the bus which is five feet in length. How can you bring the cue on to the bus without breaking either the cue or the company's rule?

ANSWERS ARE GIVEN ON PAGE 274.

Helping your child understand shapes and solids

In the primary school the treatment of shapes and solids is naturally fairly informal and exploratory. Young children aren't expected to learn geometry theorems and indeed some of what has been included in this chapter (for example Pythagoras' theorem) would be considered a bit advanced for most children of under eleven years. What they do need to do is lots of handling of boxes, balls and baked bean tins, seeing which ones roll, which don't and why. They need to spend time drawing pictures and using maps, playing with squares and triangles and fitting a variety of shapes together. Shapes can be drawn round with a pencil. Human and animal figures can be constructed using only triangles or circles. Some children spend hours with a pair of compasses constructing complex intersection circle patterns. By talking to your child about these sorts of activities and doing them with her, she will develop, in a literal sense, a *feel* for shapes and a confident mastery of the mathematical language that we use to describe them. This in turn will help her understand the use of shapes in the everyday world as well as in art and design.

Construction toys like Lego, Brio and Meccano not only give a child a better sense of space, but they also invite curiosity about the way constructions are pieced together and encourage a positive problem-solving attitude. Unfortunately, construction games tend to be stereotyped as 'boys' toys'. Given the passive nature of most of the toys which girls are given it is perhaps not surprising that these two topics – spatial maths and problem-solving – are ones in which girls seem to do consistently less well than boys. The overall underachievement of girls in maths, particularly around the ages of twelve to fifteen years, is something which concerns many parents and is the subject of chapter 21. Fortunately, there are a number of other creative and challenging pastimes which both girls and boys enjoy and which develop spatial awareness. For example:

- jigsaw puzzles
- origami – the delightful ancient Japanese art of paper folding (see reference 5, page 277)
- tangrams – the original Chinese jigsaw puzzle. To make your own set of Tans, trace the tangram square drawn below on to thick cardboard and carefully cut out the seven pieces. Now try to arrange *all* the pieces to form (without overlapping) different shapes – a bird, a rider on horseback, a cat, a triangle . . . the possibilities are endless.

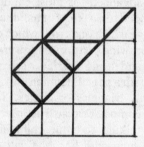

- constructing a 3D shape from a 'net' drawn on cardboard

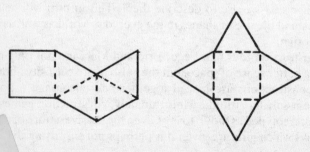

SHAPES AND SOLIDS

If these are cut round the solid lines and folded along the dotted lines, they will form a 3D shape (in this case a pyramid with a square base). Children will spend hours happily gluing these models and painting the faces. Books of nets drawn on cardboard make excellent birthday presents (see reference 6, page 277).

Finally, here is a list of the main skills to do with shapes and solids.

SHAPES AND SOLIDS CHECKLIST

- knowing the names of the common 2D and 3D shapes and their main properties (for example, that the opposite sides of a rectangle are equal and the angles are all 90°)
- knowing words like diagonal, circumference, perimeter, angle, straight, curved, sides, faces, edges, vertices
- knowing which shapes fit together and which don't
- being able to handle a ruler, compasses and a protractor. (The use of compasses and protractor has not been covered here.)

ANSWERS TO EXERCISES FOR CHAPTER 16

Exercise A

Number	6	9	4	3	7	11	12
Angle (°)	180	270	120	90	210	330	360

Exercise B

No comment.

Exercise C

No comment.

Exercise D

(i) The square of the diagonal $= 12^2 + 7^2 = 193$
Thus diagonal $= 13.9$ cm

(ii) The square of the diagonal is about 50.
Since the sides of the square are equal, the square of each side must be 25.
So the side of the square $= 5$ cm.

Exercise E

Circle	1	2	3	4	Average
C/D ratio	3.15	3.20	3.06	3.16	3.1425

This is a good classroom activity since errors of measurement tend to average out when a large number of samples are taken (as has happened here).

17
A sense of proportion

It is easy enough to distinguish children from adults. For one thing, children are much smaller. But why are we able to tell them apart in photographs? It's not just to do with skin texture, dress or body posture. Children are actually a different shape than adults. Have a look at these two outline drawings. Which one do you think represents the child and which the adult?

Although the drawings are the same *size*, the *proportions* are different. Drawing A, which represents the child, has a larger head in relation to the size of the body. The differences between adults and children in this respect are quite dramatic, as the table below shows.

Table 1 Heights and head circumferences of children and adults

Name	Age	Height (cm) H	Head circumference (cm) C	H/C
David	1 day	51	35.5	
Carrie	2 years	82	49	

Name	Age	Height (cm) H	Head circumference (cm) C	H/C
Lydia	5 years	114	52.5	
Ruth	10 years	127.5	54.5	
Luke	12 years	144	54.5	
Hilary	33 years	157.5	55.5	
Alan	37 years	172	55	

Exercise A

(i) For each of the seven people in Table 1, calculate (with the help of your calculator) their height (H) divided by their head circumference (C). Write your results in the final column of the table headed H/C.
What do these results suggest?

(ii) Now take the same measurements from members of your own family and check that the same is true for them.

COMMENTS ON THIS EXERCISE ARE GIVEN ON PAGE 205.

You may not have been consciously aware that adults and children differ so dramatically in terms of their body proportions. Perhaps one of the reasons that people have difficulty with the topic of *proportion* in mathematics is that we tend to exercise our 'sense of proportion' at an instinctive or subconscious level. Let's therefore be clear about what the word proportion really means.

What is proportion?

A common criticism of children's drawings (indeed of my own drawings!) is that certain bits are not 'in proportion'. That means

A SENSE OF PROPORTION

that they are either too big or too small in relation to the rest of the masterpiece. Imagine that you have drawn a picture of your house, reducing it in scale to one twentieth of its size. Thus:

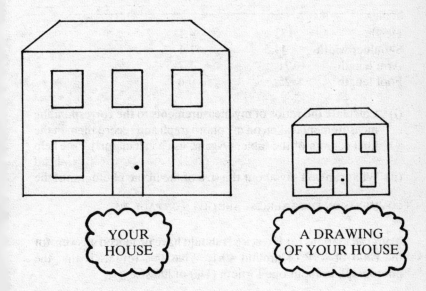

If your drawing is to be 'in proportion' then every detail must be ½₀ of the original. So if the door of your house is 2 m high, the door in your drawing should be 10 cm high if it is to be in proportion. In other words, if you take *any* measurement from your house and divide it by the corresponding measurement in your drawing, the answer should be exactly 20.

The answer that you get when you divide any two measurements is called the *ratio*. Exercise B will give you some practice at calculating ratios.

Exercise B

Table 2 below contains some of my body measurements along with the corresponding measurements taken from a photograph of myself.

Table 2

	Me (cm) M	My photograph (cm) P	Me/Photograph Ratio M/P
Height	173	4.1	
Shoulder width	44.5	1.1	
Arm length	71	1.7	
Foot length	25	0.6	

(i) Calculate the ratios of my measurements to the corresponding measurements taken on my photograph and record them in the final column of the table. (Again, use a calculator!)

(ii) What can you say about the size of me in the photograph?

COMMENTS ON THIS EXERCISE ARE GIVEN ON PAGE 205.

Your calculations for Exercise B should have produced answers for the ratio of M/P of around 40:1. What this reveals is that the photograph is about one fortieth (1/40) of life size.

Two shapes which are in proportion to each other have the same shape. In mathematics we say that they are *similar*. Unfortunately the word similar is used rather more precisely in mathematics than in everday usage. For example a teacher might be unhappy that two exam scripts look so 'similar'. Two sisters or brothers might look 'similar'. Here the word means simply 'alike in certain respects'. What it is which makes them alike is often not very clear. In mathematics, however, when we used the word similar we are referring to the same *shape*. The two triangles drawn below are said to be similar because one is an exact scaled-up version of the other.

A SENSE OF PROPORTION

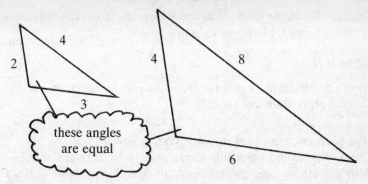

Notice also that because they are the same shape, the corresponding (matching) angles are equal.

Exercise C

Which of the following shapes are similar (in the mathematical sense)?

(i) Any two squares
(ii) Any two rectangles
(iii) Any two circles
(iv) Any two equilateral triangles (see chapter 16 if you've forgotten what these are)
(v) Any two right-angled triangles

COMMENTS ON THIS EXERCISE ARE GIVEN ON PAGE 206.

Proportion in mathematics

So far I have looked at proportion simply in terms of shape. In fact the idea of proportion in maths is much more far-reaching than this and lies at the heart of a number of mathematics topics – for example, fractions, percentages, graphs, and rates of change. Here are two examples of proportion at work in questions on number.

Example 1   How many tenths is ¾?

Example 2 In two hours I can cycle 18 miles. How far will I travel in 2½ hours at the same speed?

Exercise D

Have a go now at these two questions. As you do them try to use the ideas of proportion and ratio.

COMMENTS ON THIS EXERCISE ARE GIVEN BELOW.
The answers to these questions were respectively, 7½ and 22½ miles. Although there are formal methods for doing these sort of questions, I will now write out the solutions in such a way as to emphasize the notion of proportion.

SOLUTIONS

Example 1   ¾ = ?/10

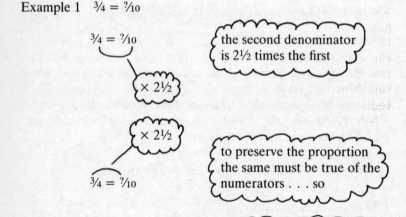

Example 2   2 hours → 18 miles
2½ hours → ?

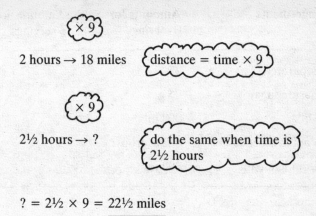

$$? = 2\tfrac{1}{2} \times 9 = 22\tfrac{1}{2} \text{ miles}$$

The idea which I find helpful in all problems on proportion is that of a *scale factor*. In the two examples above the scale factor was the number which we multiplied by (2½ and 9 respectively). With the earlier example of my photograph, the scale factor was 40. The essence of all problems to do with proportion is that, when comparing one set of numbers with another, the scale factor which connects corresponding pairs of values is always the same.

Now for a more practical example of proportion. Exercise E asks you to have a go at scaling the ingredients of a recipe.

Exercise E

The ingredients for 6 servings of hazelnut ice cream are given below. Complete the table for 8 servings.

Ingredients	Amounts for 6 servings	Amounts for 8 servings
Toasted hazelnuts		225 g
Cornflour	2 tablespoons	

Ingredients	Amounts for 6 servings	Amounts for 8 servings
Separated eggs	2	
Castor sugar	75 g	
Milk	300 ml	
Vanilla essence	a few drops	
Double cream	300 ml	

COMMENTS ON THIS EXERCISE

There is no single correct way of tackling this problem. My approach was to say that 8 is $\frac{1}{3}$ more than 6 so I increased each number in the second column by a scale factor of $1\frac{1}{3}$. In some cases this was fairly straightforward. For example:

$$\text{Hazelnuts} \quad 225 \times 1\frac{1}{3} = 300 \text{ g}$$

$$\text{Castor sugar} \quad 75 \times 1\frac{1}{3} = 100 \text{ g}$$

$$\text{Milk} \quad 300 \times 1\frac{1}{3} = 400 \text{ g}$$

However, the others were less obvious. It is difficult to crack $2\frac{2}{3}$ eggs for example! My solution would be to use 3 small eggs and 3 tablespoons of cornflour (perhaps ensuring that the third spoon would not be quite full). Of course, not every measure in a recipe needs to be scaled by the same scale factor. Take cooking time and oven temperature, for example. These are areas where the mathematics of proportion breaks down and common sense and practical experience take over.

Where proportion fits into maths

As I have already suggested, the idea of proportion is important in comparing both numbers and shapes. The tree diagram is therefore rather difficult to draw but will look something like this:

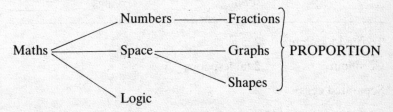

Our sense of what is beautiful in art, nature and music is closely bound up with proportion. The next chapter, on Patterns, takes a look at how mathematics can enhance our appreciation of the world around us.

PRACTICE EXERCISES

1. 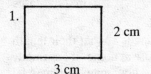 Draw a diagram the same shape as this one, but with the longest side 5 cm.

2.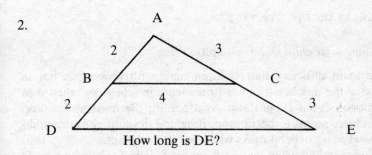

How long is DE?

3.
The ingredients for 6 servings of hazelnut ice cream are given below. Complete the table for 15 servings.

Ingredients	Amounts for 6 servings	Amounts for 15 servings
Toasted hazelnuts	225 g	
Cornflour	2 tablespoons	
Separated eggs	2	
Castor sugar	75 g	
Milk	300 ml	
Vanilla essence	a few drops	
Double cream	300 ml	

4.
Sugar lumps in cafés and restaurants are made so that their length of side is half as big as the domestic ones I use at home. If I normally take two domestic sugar lumps in my tea, how many commercial ones should I take to get the same amount of sugar?

ANSWERS ARE GIVEN ON PAGE 275.

Helping your child understand proportion

The main difficulty that children have with proportion lies in making the link between what they learn in school and their own intuitions about proportion. As a parent, the most useful task, therefore, seems to be in supporting and developing your child's natural sense of proportion whenever it is relevant.

For example, when you go shopping together to buy household items like washing powder or breakfast cereal you will need to compare the various sizes to decide which size is better value for money. (It *isn't* true that the large economy size always works out cheapest.) Another likely context for calculations involving proportion occurs when adjusting recipe ingredients for different numbers of servings. Again the idea frequently crops up in dressmaking and toy-making. Your child might be surprised to discover that the baby rabbit isn't an exact scaled-down version of the mummy rabbit. Here are one or two further situations where ideas of proportion are not very far away.

- making scale models (you may see the scale factor on the box written as 'Scale – 1:144')
- map reading – look at the bottom of the map where it should say something like 'Scale = 1:50,000'
- bicycle gears and their ratios
- any problem involving percentages or common fractions

ANSWERS TO EXERCISES FOR CHAPTER 17

Exercise A

Name	Age	H/C (2 decimal places)
David	1 day	1.44
Carrie	2 years	1.67
Lydia	5 years	2.17
Ruth	10 years	2.34
Luke	12 years	2.64
Hilary	33 years	2.84
Alan	37 years	3.13

For most adults their height is about three times their head circumference. However, with very young children their head circumference is about one and a half times as big as this in relation to their height.

Exercise B

(i) The four ratios are, roughly, 42.2, 40.5, 41.8 and 41.2.
(ii) The photograph is about one fortieth (1/40) of life size.

Exercise C

(i) Yes.
(ii) No – only if their ratios of length/breadth are the same.
(iii) Yes.
(iv) Yes.
(v) No – *all* the corresponding angles must be equal.

Exercise D

No comment.

Exercise E

No comment.

18
Patterns

How do you know when to plant your lettuce seeds? How is it that you know when your child's cough is serious? What is it about a favourite tune or picture that gives you such pleasure?

One answer to these questions is that we make sense of the world by creating *patterns* from the seeming disorder around us. The patterns we see are our way of explaining how things are. For example, after years of watching plants grow, you start to realize that there is a growing cycle between spring and autumn. Similarly, when you've heard your child cough several thousand times, you begin to develop an ear for what is serious and what isn't. Of course you don't need to check the cough against a chart to know whether she has a chest infection. Mostly the patterning we do isn't as formal as this. But there *are* times when it is helpful to be a bit more explicit about patterns. And this can even be true of something as unlikely as the appreciation of beauty. The purpose of this chapter is to show that a mathematical understanding of pattern can sometimes add to your enjoyment of art and nature. We start by looking at a mathematical pattern.

The Fibonacci Sequence

Fibonacci was an Italian mathematician who gave his name to this interesting series of numbers. See if you can work out the next two in the sequence.

1, 1, 2, 3, 5, 8, 13, ☐, ☐ . . .

Did you spot that each number is the sum of the two previous numbers? So 13 was found by adding the 5 and 8. The next two numbers are therefore *21* (from 13 + 8) and *34* (from 21 + 13).

What makes this pattern particularly interesting is the fact that it often appears in nature, describing, for example, the spacing of the leaves of certain plants and the arrangement of petals and thorns. The next time you pass a hawthorn hedge, look at the arrangement of the thorns on each twig.

A HAWTHORN TWIG

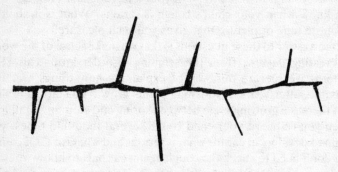

You'll find that the thorns always rotate *2* times round the stem while the pattern repeats every *5* thorns. Both the numbers 2 and 5 can be found in the Fibonacci sequence. More remarkable is that when certain other plants are analysed in this way, the two numbers which describe their repeating pattern are always Fibonacci numbers. Here are some examples:

Table 1

Plants	Rotations	Repeating pattern
Hawthorn, apple, oak	2	5
Beech, hazel	1	3

	PATTERNS	
Poplar, pear	3	8
Leek, willow, almond	5	13

Exercise A

Make a note in your diary for next weekend to examine an orchard or hedgerow so that you can see some of these patterns for yourself.

Not all plants have thorns but many others have a combination of clockwise and anticlockwise spirals at their centre. Again Fibonacci numbers usually describe the number of spirals in each direction. For example:

        Daisy   (21, 34)
        Sunflower   (55, 89) or perhaps (89, 144)

Curiouser and curiouser.
   Now let's look at another remarkable pattern.

The Golden Ratio

It is very difficult to explain why we find some shapes more attractive than others. Yet it does happen that many people have similar preferences. Perhaps there *is* an underlying explanation but we are not aware of it. Try out your own preferences now in Exercise B.

Exercise B

Look at the rectangles and crosses below.
Choose the one from each which seems most 'right' to you.

Rectangles

(a) (b) (c)

Crosses

(a) (b) (c)

COMMENTS ON THIS EXERCISE

I wonder if your choices were in line with most people's that rectangle (a) and cross (b) had the right sort of 'feel' to them. The thing that gives a particular rectangle its *shape* is the contrast between its length and its breadth. If you measure the dimensions of rectangle (a) above, you will find that its length is about 1.6 times its breadth. Similarly, the upright part of cross (b) has been divided into two portions which are also in the same proportion – that is, the lower portion is about 1.6 times as long as the upper portion. The shorthand way of describing this proportion is to say that the upright is divided *in the ratio* 1.6 to 1. Similarly, the sides of rectangle (a) are in the ratio 1.6 to 1.

The ratio of about 1.6 to 1 is known as the 'golden ratio' because it has certain special properties. (As you will see shortly, its exact value is slightly more than 1.6.) To understand what is so special about this ratio, look at the diagram below:

The line AB is 2.6 units long. The point P divides the line in the ratio 1.6 to 1. Now do Exercise C.

Exercise C

(i) Use your calculator to work out AP/PB and AB/AP.
   AP/AB = _____ AB/AP = _____

   What do you notice about your answers? _____

(ii) Now try the same exercise with a slightly larger value for AP (say 1.62).
   What do you notice? _____

   _____

COMMENTS ON THIS EXERCISE ARE GIVEN ON PAGE 218.

If you keep experimenting with the length of AP you will find that the values of the two ratios get closer and closer together as AP gets closer to the true value of the golden ratio. This number is a bit like π in that you can never quite get it exactly. However, here are its first eight digits.

   Golden ratio = 1.6180340 . . .

Now pick up your calculator again and, using the values AP = 1.6180340 and PB = 1, calculate the ratios AP/PB and AB/AP. You should find that, within the limits of accuracy of the calculator, the two ratios are equal.

The golden ratio has been given a special letter, the Greek letter ϕ or phi. The proportion has been known and used in paintings, sculpture and architecture since the times of the early Greeks (the Parthenon is a good example). In 1509, Leonardo da Vinci illustrated a book called *De Divina Proportioni* (*The Divine Proportion*) showing this golden ratio in various shapes and solids. Indeed, over the centuries people have thought that this particular proportion had magic properties. Even today it is all around us. Try measuring the proportions of playing cards, books, windows and picture frames. You'll be surprised how many of these rectangles have their sides close to the proportions 1.618 to 1. Very mysterious!

Exercise D

(i) What sort of answers do you think you would get if you went through the Fibonacci sequence dividing each number into its successor? Using a calculator, fill in the table below and find out.

Fibonacci ratio	1/1	2/1	3/2	5/3	8/5	13/8	21/13	34/21	55/34
Decimal value	1	2	1.5						

(ii) Check whether this is a golden rectangle. → [rectangle with sides 5 and 8.09]

(iii) With the help of a calculator, find the value of $1/\phi$. What do you notice about your answer?

(iv) Divide your height (from head to toe) by the height of your navel (from navel to toe). (You might even like to survey your adult women friends on this one.) In theory this should give you a value very close to ϕ.

COMMENTS ON THIS EXERCISE ARE GIVEN ON PAGE 218.

Patchwork Quilts

Have you ever tried to make a patchwork quilt? If so you have probably discovered for yourself that it isn't always easy to make shapes fit together. For example, anything with curves, like circles, is hopeless. Squares and rectangles on the other hand fit well together but give a rather boring overall effect. The really interesting designs have a more unusual basic shape – perhaps a hexagon:

or a combination of squares and triangles:

In mathematics, the word for fitting shapes together so that there are no gaps or overlaps is *tessellation*. Many modern designs of wallpaper and fabrics are based on tessellating shapes. Hundreds of years ago, Moorish mosaics were built up from highly complex tessellations.

Part of a mosaic from a thirteenth-century Moorish palace in Spain.

From *The Search for Solutions,* Horace Freeland Judson, Hutchinson, 1980, page 40.

More recently a Dutch mathematician and artist, Maurits Escher, has produced a number of breathtaking drawings and engravings which are based on tessellating shapes – including birds, fish and other creatures from his imagination.

From *The Graphic Work of M. C. Escher* (Pan/Ballantine, 1972)

PATTERNS

Tessellating patterns appear on a vast scale in nature. Keep a look out for some of these below. They might inspire your next patchwork quilt:

- markings on a giraffe or a turtle shell
- cracking in solids like ice, lava or clay (e.g., the Giant's Causeway in Northern Ireland)
- clusters of bubbles
- honeycombs

Exercise E

Try to tessellate the following shapes:

(i) An equi-lateral triangle

(ii) An isosceles triangle

(iii) A scalene triangle

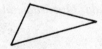

(iv) An irregular quadri-lateral

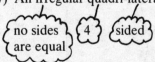

(v) A regular penta-gon

(vi) A regular hexa-gon

COMMENTS ON THIS EXERCISE

You may be surprised to discover that all of these shapes, apart from the regular pentagon, do tessellate. With some you will have to shift the pieces around. For example the irregular quadrilateral (iv) will tessellate by putting equal sides together. If you cut a stencil out of a piece of cardboard, like this, you can tessellate the shape thus:

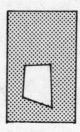

An Irregular Quadrilateral

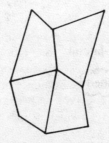

Children and Patterns

As a child, although I enjoyed painting and drawing pictures, I never found it particularly easy. I didn't seem to possess a 'natural eye' for proportion, or, if I did, I was unable to translate it on to paper. What *did* give me enormous satisfaction was the more geometric constructions which could be carefully ruled and coloured. For most children constructing geometric patterns is an activity which they never seem to tire of. Christmas is an ideal opportunity for children to design their own cards and decorations using tessellating shapes. Other activities already mentioned in this chapter are constructing mosaics and patchwork quilts. Next time you and your child are out of doors together, take a really close look at some of the wonderful patterns in nature – a spider's web, a leaf, a honeycomb, a flower, the pattern of branches on various trees, a snowflake, cracks in mud or ice, animal markings. . . . Have a look also at the variety of patterns of bricks in walls and buildings. Your child may be interested to discover the many tessellating patterns on fabrics, wallpaper and posters. She may even be moved to attempt some of her own drawings in the style of Escher (see page 214). For any pattern you choose to explore, here are two questions which may help provoke your child's curiosity and interest.

- *What* is the pattern? – How would you describe it?

and ● *Why* is the pattern as it is?

Summary

This chapter concentrated on two popular mathematical patterns which are closely connected – the Fibonacci sequence and the golden ratio. The first of these appears often in nature, while the second has had a considerable influence on art and architecture for thousands of years. Finally we looked at patterns in patchwork quilts and tried to discover which shapes would and would not fit together.

```
                    Numbers
                   /        \
        Maths ——— Space ———— Pattern
                   \
                    Logic
```

ANSWERS TO EXERCISES FOR CHAPTER 18

Exercise A
No comment.

Exercise B
No comment.

Exercise C
(i) The two ratios AP/AB and AB/AP are very nearly the same.

(ii) When AP = 1.62, the two ratios are even closer in value to each other.

Exercise D
(i)

Fibonacci ratio	1/1	2/1	3/2	5/3	8/5	13/8
Decimal value	1	2	1.5	1.6666667	1.6	1.625

Fibonacci ratio	21/13	34/21	55/34
Decimal value	1.6153846	1.6190476	1.6176471

(ii) The ratio of the sides = 8.09:5
 = 1.618
As you can see, this is very close to the value of φ.

(iii) On the calculator I press:

1 ÷ 1.61803398 =

Your calculator may have given a slightly different answer – mine rounded up the sixth decimal place. What is interesting about the answer is that the decimal part is exactly the same as the decimal part of φ. Phi is the only number which has this property.

19
Diagrams and graphs

It is difficult to open a newspaper or watch television without hearing a statistical fact. For example, did you know that eight out of ten advertisers are prepared to mislead the public a little in order to sell their product? (Incidentally, the other two are prepared to mislead the public a lot!)

Sometimes, as with my example above, statistical information can be based on pretty thin evidence. However, what is more likely to upset our notions of 'honest, truthful and legal' is the way in which statistics are presented to the public, for here there is much scope for deception. Graphs and diagrams, therefore, are not only a useful and important topic for children in their own right, but they also provide interesting insights into the way facts can be distorted and the public duped.

Diagrams

Diagrams are usually drawn as an aid to thinking. They are a useful way of sorting out your ideas and clarifying your thoughts on paper. Here are a few of the common diagrams children meet in school.

A. *Venn diagrams*

Children use Venn diagrams to help them sort or classify various things into sets. For example:

Squares   Triangles

DIAGRAMS AND GRAPHS

Here is a more complicated Venn diagram, this time with *intersecting* circles.

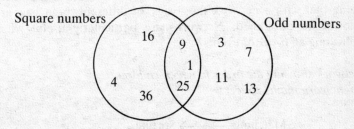

B. *Tree diagrams*

Like Venn diagrams, tree diagrams are useful to help you *classify* or see how things are connected to each other. For example:

Tree diagram showing words which describe things we eat and drink

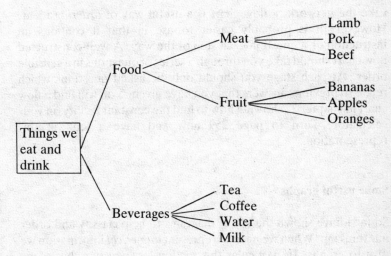

C. Network

A network diagram is similar to a tree diagram, the main difference being that with a network the 'twigs' of different 'branches' are allowed to be joined. Networks are particularly useful when indicating an *order* of events.

Network showing the order in which children might learn maths concepts

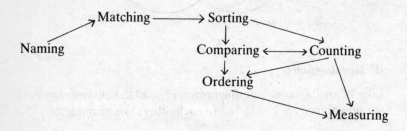

D. Flow chart

Like the network, a flow chart is a useful way of *ordering* ideas. However, it is probably easier to use in that it contains an instruction or a question each step of the way. A well-constructed flow chart should take you through a series of questions in a sensible order. At each stage you should only be asked questions which relate to previous answers that you have given. You will find a flow chart in chapter 22 showing how to find the constant facility on your calculator. Turn to page 259 now and have a look at this representation.

Some useful graphs

So far I have shown the use of diagrams to help classify and order our thinking. When we need to represent *numerical* information we turn to graphs. In particular this section looks at bar charts, pie charts, line graphs and scatter graphs.

DIAGRAMS AND GRAPHS

E. *Bar charts and pie charts*

These representations are useful when we want to make *comparisons*. Bar charts (sometimes called block graphs) are simply a set of vertical bars set side by side. The height of each bar is an indication of its size. With pie charts the size of each item is represented by the size of its slice on a pie.

Bar chart showing why men like marriage

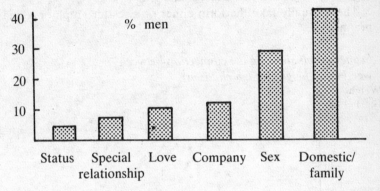

Pie chart showing why men like marriage

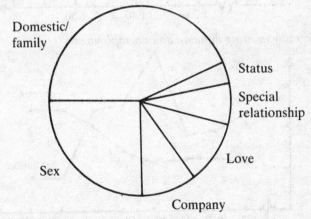

Source: *The Hite Report on Male Sexuality*, Shere Hite (Macdonald, 1981)

F. *Scatter graphs and line graphs*

Bar charts and pie charts are useful for comparing things. Sometimes, however, we need to know how two different measures are related to each other. For example, how does *height* relate to *weight*? How much has a particular plant *grown* over *time*? Is a child's *health* linked to *poverty* in the family?, and so on. To answer these sort of questions, which look at two different measures together, we need a two-dimensional graph.

These usually take the form either of a scatter graph or a line graph.

Scatter graph showing the connection between weight and height (of ten children)

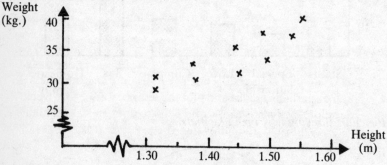

Line graph showing inflation and unemployment (UK)

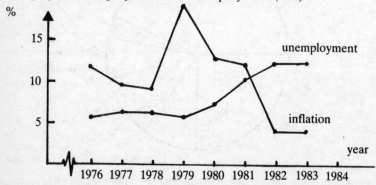

Both these sorts of graphs contain *axes* – the straight lines running up and across the page which have scales and hence enable you to mark out positions on the graph.

Helping your child with diagrams and graphs

For most children it is not until they use maths in their everyday life, and see that it works, that they fully understand and believe in the maths they are taught at school. Unfortunately, the drawing of graphs and diagrams in maths lessons does not often come to life outside the classroom. As a parent you can help your child see the relevance of diagrams and graphs by drawing her attention to the occasional representation when it crops up in a newspaper or on TV.

Two aspects in particular are worth stressing:

(a) the *purpose* of the graph or diagram – what is it designed to suggest and why has *that* particular representation been used? You could plot your child's height or weight every month on a graph or perhaps record and plot the temperature every week. One helpful activity is to ask your child to discuss the 'story line' in time graphs like the following:

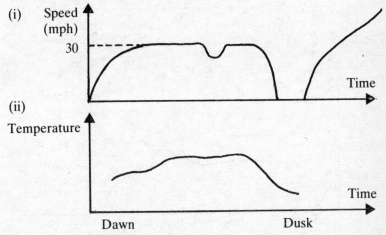

(b) encourage your child to play detective in spotting misleading representations which are designed to create a false impression. Newspapers and magazines are a rich source of misleading graphs. Things to watch out for are the following:

- no title or labels on the axes
- no scale marked on the axes
- the scale on either axis not starting at zero
  (in fact this is acceptable *provided* an indication is made that this has been done – for example a break in the axis, thus —⋀⋁⋀—) 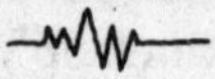
- the sort of misleading graph shown on page 172.

20
The X factor – algebra

For most of us, algebra first appeared on the scene early on in the secondary school. It was, perhaps, the beginning of the end as far as maths was concerned. Although it is still a secondary topic, it is included here to let you get one step ahead and to anticipate in advance what your child will soon be doing. Also, if you didn't understand algebra at school you might just have a nagging curiosity as to what it was all about.

Algebra is sometimes described as the study of the twenty-fourth letter of the alphabet. For pupils used to working with numbers in maths lessons, it comes as a bit of a shock to be asked to play around with letters, like x and y. There seems to be no point to it – just a strange game with seemingly arbitrary rules.

Well, school algebra can seem pretty pointless. Most children never get beyond struggling with the symbols and conventions and therefore never experience for themselves the power of algebra to clarify and solve problems. The aim of this chapter is to give a flavour of what algebra can do, rather than listing the rules and conventions. Let's start with a problem involving numbers.

Look at this sequence of numbers and then try the exercise below:

3, 5, 7, 9, 11 . . .

Exercise A

Write down:

(i) the next number in the sequence
(ii) the tenth number in the sequence

(iii) the hundredth number in the sequence

COMMENTS ON THIS EXERCISE
Exercise A is concerned with spotting number patterns and therefore seems to be an exercise in arithmetic rather than algebra. This may be true for your answers to parts (i) and (ii) (respectively 13 and 21), but how did you do part (iii)? If you got the correct answer of 201, I simply don't believe that you actually counted out one hundred numbers starting with 3, 5 . . . and so on. What is more likely is that you tried to find a pattern in the numbers. Exercise B asks you to think about how you (might have?) answered this question.

Exercise B

The pattern which we are looking for here is the connection between the *position* of each number in the sequence and its *value*. Look at the following table and then try to complete the sentence below.

Number *position*	1 2 3 4 . . . 10 . . . 100 . . .
Number *value*	3 5 7 9 . . . 21 . . . 201 . . .

To get the *value* of a number you take its position in the sequence _____

COMMENTS ON THIS EXERCISE
I completed the sentence in Exercise B as follows:

'. . . multiply it by 2 and then add 1'.

Let's take an example from the table and see if my formula works. The number 11 is in position 5.

Yes, 11 *is* equal to $2 \times 5 + 1$.

THE X FACTOR – ALGEBRA

Here, then, is the formula for finding the value of any number in the sequence 3, 5, 7, 9, 11 . . . , knowing its position in the sequence.

> Multiply the 'position number' by 2 and add 1

Now we have moved from arithmetic to algebra! Whereas arithmetic was concerned with *particular* numbers, algebra expresses a *general rule for any number*. We still haven't started to use symbols like x and y, but that can come later. First you'll need some practice at moving from the particular to the general, so do Exercise C now.

Exercise C

Below you will find some important properties of arithmetic expressed with particular numbers as examples. Write in your own words the general rule in each case. The first one has been done for you.

Example	General rule
(i) $4 + 3 = 3 + 4$	You can add two numbers in any order and still get the same answer.

(ii) $5 + 5 = 2 \times 5$

(iii) $2 - 5 \neq 5 - 2$

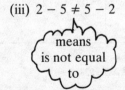

means is not equal to

COMMENTS ON THIS EXERCISE ARE GIVEN ON PAGE 237.

Introducing x and y

Having got the main purpose of algebra safely out of the way (algebra is a way of stating *general* rules), we can now look at the way we write it. The second key feature of algebra is that instead of writing sentences in words, we have invented a series of abbreviations to speed things up. I know that to most people the symbols in algebra seem to cause confusion rather than be an aid to efficiency. However the idea isn't just confined to maths. In any activity where people want to get across a lot of information in a short space, they invent a shorthand. Look at some of the examples in Exercise D for instance.

Exercise D

Can you spot the source of these shorthands and what they mean?

Example	Source	Meaning
(i) Det. hse, lge gdns, gd decs, FCH . . .		
(ii) K1, P1, M1, C4F, K2		
(iii) NYWJM seeks same with view to B & D, S & M, etc.		

COMMENTS ON THIS EXERCISE ARE GIVEN ON PAGE 238.

The main abbreviation in algebra is to use just one letter to represent the thing whose values we are interested in. Here is an example:

THE X FACTOR – ALGEBRA

In words	*In algebra*
My weekly milk bill in pence is equal to 23 times the number of pints I buy (23p being the cost of a pint of milk).	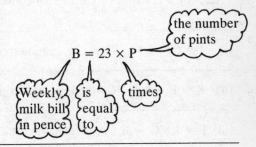

Not only have I introduced the abbreviations B and P, but I have also included the two useful symbols = and × to save me the trouble of writing out the words 'equals' and 'times'. As you see, I have had to be quite clear exactly what each letter stands for, and that includes stating the units in which they are measured (pence and pints).

The letters in algebra which we happen to choose to represent numbers are quite arbitrary. It is usually a good idea to relate the letter to the quantity that it represents. For that reason I used the letter B to represent the *b*ill and P for the number of *p*ints. However, the most common letters used at school level are:

x, y, a, b and n.

There seems to be no good reason for this choice, with the possible exception of the n, which can be thought of as representing some unknown *n*umber.

Although a formula can be written out in words, it is usual to express it in this algebraic shorthand. For example, here is a simple formula:

$$F = 3Y$$

This formula converts the number of yards (Y) into feet (F). So if you wished to know how many feet there were in 8 yards, simply replace the letter Y by 8. Be careful, however, for the answer is *not*

38! In maths 3Y is taken to mean 3 times Y. So the answer should be 3 *times* 8, or 24 feet.

Here are a few other formulae for converting units.

(i)   I = C ÷ 2.54   Converts centimetres (C) to inches (I)

(ii)   K = P ÷ 2.2   Converts pounds weight (P) to kilos (K)

(iii)   F = 1.8 C + 32 Converts temperature from degrees Celsius (C) to degrees Fahrenheit (F)

Some conventions in algebra

If we wished to write 'x times y', it would look like this:

x × y

As you can see there is a real danger of confusing the letter x with the multiplication sign ×. This problem is solved in algebra by dropping the multiplication sign altogether.

Thus:

xy	means	x times y
2x	means	2 lots of x or 2 times x
3a + 2b	means	3 times a plus 2 times b

(i.e., add)

Another common convention is the use of brackets.

4(x + 2y) means 4 *times* what is in the brackets – i.e., the x + 2y

In other words, *everything* inside the brackets is multiplied by the number immediately outside it (giving the value 4x + 8y).

A third important convention is used when a number or letter is multiplied by itself. For example 3 × 3 is written as 3^2 (three squared).

THE X FACTOR – ALGEBRA

Similarly:

x × x	is	x^2 (x squared)
3 × y × y	is	$3y^2$ (three y squared)
and y × y × y	is	y^3 (y cubed)

You'll need some practice at using these conventions, so try Exercise E now.

Exercise E

If a = 6, b = 4 and c = 3, find the values of the following:

(i)   a + b
(ii) a − b + c
(iii) a + 2b
(iv) 2a − bc
(v) $(a + b)^2$
(vi) a(2b + c)
(vii) $2c(a^2 − b^2)$
(viii) $(a − b)^2 − c^2$

ANSWERS TO THIS EXERCISE ARE GIVEN ON PAGE 238.

What is the value of 2 + 3 × 4?

Most people would give the answer 20, since 2 + 3 gives 5 and 5 × 4 gives 20.

Now try this one:

What is the value of 2 + 3x   when x = 4?

When you give x its value of 4, the expression again becomes 2 + 3 × 4. However, perhaps you might be unhappy about giving an answer of 20 here. You may see that another way of calculating the value of the expression is possible, giving an answer of 14. This is done by multiplying 3 × 4, giving 12, and then adding 2.

Now try the sequence 2 [+] 3 [×] 4 [=] on your calculator. You may be surprised to discover that some calculators (generally the more basic ones) give the answer 20 while others give 14. Calculators which give 14 are commonly known as 'algebraic' since they obey an important algebraic convention that when there is a choice, [×] and [÷] should be carried out before [+] and [−]. Here are one or two further examples.

	Answers	
Key sequence	Arithmetic	Algebraic
2 × 3 + 4 =	10	10
2 + 3 × 4 =	20	14
6 − 4 ÷ 2 =	1	4
6 ÷ 2 − 4 =	−1	−1

Proving with algebra

This final section looks at an aspect of maths close to the hearts of mathematicians – the idea of *proof*. Without algebra, proving that a mathematical result is true is quite difficult. It is easy enough to show that the result is true for several particular numbers but it is quite a different matter to say that you *know* it is true for *all possible numbers*. Have a look at this 'Think of a Number' exercise.

Exercise F   Think of a number

Have a go at the following:

   Think of a number between 1 and 10.
   Add 2.
   Multiply by 3.
   Double it.
   Divide by 6.
   Subtract the number you first started with.
   The answer is 2!

If you didn't get the answer 2, then you made a mistake in your calculation.

COMMENTS ON THIS EXERCISE
You may be wondering how I knew that the answer would be equal

to 2. In fact, no matter what number you choose to start with, the answer is always equal to 2. One neat way of proving that this is a fact is with algebra. Instead of choosing a *particular* starting number, let's start with a *general* number called x. Here are the six steps of the exercise again, based on x as the starting number.

Think of a number.	x
Add 2.	$x + 2$
Multiply by 3.	$3x + 6$
Double it.	$6x + 12$
Divide by 6.	$x + 2$
Subtract the number you first started with.	2

Now let's turn to some of the experiences that children have with algebra.

Children learning algebra

Happy Birthday to you
Happy Birthday to you
Happy Birthday dear *somebody*
Happy Birthday to you.

The singer/composer of the above is Timmy aged two. This is hardly complex algebra but in a way Timmy has discovered one of the most important algebraic ideas – that certain words (later merely letters) can be used to represent a variety of other things. For example, the word 'somebody' is a *general* word which can be used to represent various *particular* names (Richard, Julie, Oliver, and so on).

Last Christmas my own children used this idea to solve the problem of the dreaded Christmas thank-you letters. They wrote out a basic letter as follows:

Dear X,
Thank you very much for the lovely Y you sent me. We have had a very nice Christmas. . . .

Then they went through their thank-you list and inserted someone's name for the X, their gift for the Y, and so on.

In the above example, Timmy, Luke and Ruth were developing their grasp of the link between the general and the particular. Here are a few more examples of words which are general descriptions that children will come across.

General description	Letter	Some particular examples
Speed	S	30 mph, 110 kmph
Price (of a bar of chocolate)	P	16 pence, 18 pence
Number of matches in a box	M	51, 56, 47

You will notice that in the case of these last three examples the 'particular' values are all *numbers*. This is the way that letters are used in algebra. It is an important and very difficult idea for children, who find it hard to think of 'P' both as a letter and as a sort of 'empty box' which is used to represent some unknown number.

One of the ways of introducing children to using letters in algebra is to talk about apples and bananas; then later simply a and b. This 'fruit bowl' approach is quite a useful way of exploring some of the conventions of algebra (for example, you can't add 2a and 3b together because you can't add apples to bananas). However, it is actually quite confusing in another respect. If children think of the expression:

$2a + 3b$

as representing 2 apples + 3 bananas, they are misunderstanding the way in which letters are used in algebra, where *the letters themselves represent numbers*. Thus 2a means 2 times some unknown number called a. 3b means 3 times some unknown number called b. It is this main notion that *letters represent numbers* which children tend to lose sight of.

THE X FACTOR – ALGEBRA

Finally, here are one or two algebraic activities that you might like to try with your child.

(i) Play 'Think of a Number' games
   e.g., Think of a number between 1 and 10.
       Add 1.
       Multiply by 4.
       Double it.
       Divide by 8.
       Subtract the number you first started with.
       The answer is 1!

ANSWERS TO EXERCISES FOR CHAPTER 20

Exercise A

No comment.

Exercise B

No comment.

Exercise C

Example	General rule
(ii) $5 + 5 = 2 \times 5$	Adding a number to itself is the same as multiplying the number by 2.
(iii) $2 - 5 \neq 5 - 2$	Order matters in subtraction.

Exercise D

(i) Det. hse, lge gdns., gd decs, FCH . . .	Newspaper ad. for a house	Detached house, large gardens, good decorations, full central heating . . .
(ii) K1, P1, M1, C4F, K2	Knitting	Knit 1, Purl 1, Make 1, Cable 4 forward, Knit 2.
(iii) NYWJM . . .	Personal ad. in US newspaper	New York white Jewish male . . . Bondage & Discipline, Sadism & Masochism, etc.

Exercise E

(i) 10 (ii) 5 (iii) 14 (iv) 0
(v) 100 (vi) 66 (vii) 120 (viii) −5

Exercise F

No comment.

21
She who dares, wins!

CHEATING

Working together isn't cheating.
Using a calculator isn't cheating.
Finding out the answer from the back of the book
    and trying to work out how they got it isn't cheating.
Cheating is pretending you understand when you don't.
That's when you're cheating yourself.

ACCESS Department (1981)
Tower Hamlets Institute of Adult Education

Maths, particularly in the secondary school, can be hard, frustrating and at times boring. It is a subject in which you need to be prepared to 'have a go' if you are to succeed. This means being wrong sometimes. But it also means not minding too much about being wrong and being prepared to learn from your mistakes. Girls, like boys, need to risk failure in order to be successful at maths. She who dares, wins!

Jane is just fifteen and better than average at most school subjects. She plans to take five 'O' levels and three CSEs next year. Two of the CSEs are maths and science. If you ask Jane about her maths she will tell you, 'I'm hopeless at maths. I just don't understand it.' Yet four years ago in her primary school, Jane was at the top of her class at maths and really enjoyed the subject. So what has happened in the last four years to put Jane off? Later in this chapter I'll try to find some answers to this question. First, however, let's look at public examination results in maths and see if Jane is typical of other girls.

Maths examinations

Public examinations in the form of CSE (Certificate of Secondary Education) and 'O' level take place at around sixteen years of age. Although roughly equal numbers of girls and boys sit CSE maths, the more highly prized 'O' level attracts more boys (roughly speaking, for every four girls who enter 'O' level maths there are five boys). Not only do more boys try for 'O' level maths but when it comes to results they do slightly better. This is particularly noticeable if you look at the top grade in 'O' level maths which is usually achieved by about twice as many boys as girls.

The next hurdle in the examination race is 'A' level, taken at around eighteen years of age. 'A' level maths classes tend to contain only pupils who have done well in 'O' level maths and therefore have a higher proportion of boys. However, even taking this into account there are a disappointing number of girls studying for 'A' level maths. Roughly three times as many boys as girls leave school with an 'A' level in maths. As you might expect, the gender gap continues at university level and is reflected in the proportion of women maths graduates entering the teaching profession. (At present roughly one in three of the maths graduates in schools is a woman.)

The maths topics which girls find difficult

Most secondary teachers will have noticed that girls' failure in maths is not evident right across the board in all topics. A number of researchers have analysed examination performances question by question to see if there are any patterns in the differences between girls and boys. There are, in fact, topics where girls seem to outshine the boys; in particular the sort of questions which ask for a definition or require the use of a well-practised method of solution or the substituting of numbers into an algebraic expression. Sadly, these are all rather dull, unimaginative topics which can be mastered by drilling. Boys tend to do better at problem-solving and geometry questions which demand a degree of 'spatial visualizing'. This means being able to see 3D shapes and positions in your mind. In general, then, girls perform better at tests of computation (straight arithmetic), while boys seem to favour mathematical reasoning.

These findings all seem to confirm the popular stereotype that girls are just not as good as boys at maths – particularly 'hard maths' which demands logical reasoning. However, this would *not* be a logical or reasoned conclusion from the evidence presented. The point is that performance in maths exams and basic ability in maths *are two quite different things*.

Nurture or nature

It is popularly believed by some that girls do less well than boys in maths because of their female biology. Various theories about girls' genes or their hormones or the shape of their brains have been suggested to back up the claim that girls' failure at maths can be put down to their *nature*. However, there are other possible explanations. First of all it is important to realize that the gender gap in maths doesn't seem to appear till around the age of twelve to thirteen years. Up until then girls are at least as good as boys. Something seems to happen around the time of puberty which undermines girls' enthusiasm for and confidence in maths. As I will show in this chapter, there are a number of subtle (and some not so subtle) influences at work which have nothing to do with girls' nature but are rather their nurture – i.e., what they pick up from the world around them.

Why do girls learn to fail at maths?

No woman is an island, entire of herself. She is a member of a complex society which both reflects and creates the sort of person she is. Particularly around adolescence the average twelve- to fourteen-year-old girl is struggling to establish her identity as a woman. Her image of what it is to be a woman is moulded by the attitudes of her parents, friends and teachers, by what she sees and hears on TV, radio and the cinema and by what she reads in magazines, novels and school textbooks. In many respects the influences she receives are conflicting. However, in terms of attitudes to maths, society seems to speak with one voice. The essential message seems to be that it is somehow unfeminine to be good at maths. The few girls who are successful tend to be stereotyped as flat-chested intellectuals and therefore of no interest whatever to boys.

In the next two sections I will look at some specific factors which may directly or indirectly affect girls' attitudes to maths. The first of these is books and magazines.

The influence of books and magazines

I have in front of me a Ladybird book for toddlers, a children's comic and a school maths textbook. In different ways I feel that they all undermine girls' attitudes to maths. Let's look first at *Things We Like*, which is one of the 'Ladybird Key Words Reading Scheme' series. Here we follow the exploits of Peter and Jane. First we see them on the swings with the picture showing Peter swinging higher than Jane.

'I like this,' says Peter. 'It is fun.'

Then later on the beach,

'Get in the boat, Jane,' says Peter.

and then

'Look at me in the boat, Jane.'

Later, after Jane has offered Peter a cake, they stand at the railway station looking at the trains.

'I see the train, Jane. I like trains.'

Jane seems lost for words at this point.

Back home again, and Peter and Jane play with the toy train and toy station.

Jane says, 'Please can I play?'

'Yes,' says Peter. 'I have the train. You play with the station.'

And so it goes on and on. . . . Peter continues to grab the limelight and the active decision-making role. Jane looks on supportively and plays second fiddle. Well, what has all this got to do with maths? Quite a lot, I think, but mostly at a fairly subtle level. Jane is learning to keep a low profile. She will work hard at maths in primary school and get good marks by being neat and accurate. But unless she is prepared to break the mould of being quiet and non-disruptive – to ask challenging questions and to risk giving incorrect answers in class – her curiosity about mathematical things will dull and she will learn to settle for second best.

My second extract is taken from the comic *School Fun* (7 January 1984) and is called 'School Belle'. You can read it now on page 248–9 and draw your own conclusions.

Finally to the mathematics textbook. I have chosen at random a popular modern book. It is from the series 'Primary Mathematics: a development through activity', written by the Scottish Primary Maths Group (Heinemann Educational Books; 1984 edition). I have gone through this 39-page school text making a note of the gender of the characters mentioned in the questions and the illustrative cartoons which it contains. The results are summarized below.

Table 1

	Characters	Cartoons
Number of boys	10	20
Number of girls	4	13
Number of men	0	2
Number of women	0	1

Most maths textbooks are written by men and this perhaps accounts for the bias of male characters and interests (questions on cricket averages and football scores, for example). It isn't difficult to find a maths textbook with more male bias than this. At the more extreme end of the spectrum is *Concise Modern Mathematics* by D. G. Munir (Longcress Press, 1978). The first hundred people mentioned in the problems questions in this text break down as follows:

Number of men	41
Number of women	2
Number of boys	53
Number of girls	4

In this particular author's mind, women and girls seem to be almost, but not quite, totally invisible. On the rare occasions when they do appear, girls are put firmly in their place. For example:

In a hundred yards' race a boy beats a girl by 12 yards.
How far has the girl run when the boy has finished?

and another question, which begins thus:

D = (girls who play with dolls), s = (girls who like sewing). If D . . .

The problem, then, is not simply one of quantity – the number of occasions that girls get a mention in these texts. Just as important is the sort of girl that is portrayed. Too often she is a traditional, passive, 1950s, cardboard cutout who has sweets shared out for her and rail tickets bought for her – not at all the type that my daughter, for one, would identify with! Fortunately, however, this state of affairs has been acknowledged to be damaging to girls and, increasingly, publishers are ensuring a better gender balance in their classroom materials. Next time you pick up your child's maths textbook you might like to do a similar analysis. You could even show the results to your child's teacher!

Other influences

Do you remember maths lessons as opportunities for mutual cooperation and support, or were they more individual and competitive in style? Certainly in most secondary schools the latter approach seems to be favoured by maths teachers. Perhaps this is partly due to teachers' understandable wish to keep a tight grip on the discipline of their class. They feel it is easier to control a class where all communication is channelled through the teacher, than if pupils are discussing and exploring ideas in groups of two or three. Whatever your views on the link between teaching style and discipline, there is evidence that girls work better in an atmosphere of cooperation and prefer learning quietly in their own time from a classmate rather than under the public eye of the teacher.

Careers are another factor influencing attitudes to maths in school. Many girls give up the struggle with maths at around the age of fourteen, thinking that they will never need a qualification in the subject when they leave school. As school leavers look ahead to continuing bleak career prospects (combined with rapid changes in the sort of jobs being offered), too many girls are closing off what few options they do have.

Computers are making a significant and growing impact in what happens in school. It is clear that in just a few years, computer-based methods of learning will be firmly established, particularly with subjects like maths. However, already girls are opting out. At present for every girl who uses a computer at home there are thirteen boys. In the school computer clubs (held at lunchtimes and after school) the girls are similarly outnumbered. If this trend continues the 'maths gap' between girls and boys will steadily widen.

Helping girls succeed

In recent years girls' underachievement in maths has become a major issue amongst teachers and educationalists. Much of the initial effort has gone into raising teachers' and parents' awareness of the extent of the problem and trying to understand the many factors which undermine girls' confidence in maths. More recently a number of teachers have felt that 'consciousness-raising is all very well but when are we going to take positive action to redress the balance?' Some initiatives have been tried with single-sex streaming for mathematics only. Although at first the girls-only maths classes seemed to do noticeably better than those taught with the boys, it is now not clear whether this finding is consistent between different schools and over the long term.

How can I encourage my daughter with maths?

It is one of those unfortunate quirks of timing that the arrival of adolescence coincides with the preparation and taking of public exams. As a result, worries about percentages and algebra get bound up with acne, friendships and popularity. Adolescence is a time of conflict and paradox – you model yourself on your parents, yet reject everything they say and do. It is a time when parents ritually hand out 'good advice' and their children ritually ignore it. The possibility is, therefore, that your efforts to encourage your daughter's maths will be misunderstood or rejected out of hand ('My mum thinks I'm better at maths than I am'). My advice, therefore, is to provide what support you can *right from the beginning*. This means stimulating her curiosity and building her self-confidence right through the primary years. The following practical tips are relevant to both mothers and fathers. However, given the importance of the mother-daughter relationship in providing a role-model, I would suggest that mum might take most of the initiative here.

- See her teacher regularly and show as much interest in her progress in maths as you would with your son's teacher.
- Talk to her about various jobs and broaden her horizons and career ambitions.
- Encourage her to ask questions at home and at school and to expect that her ideas will be taken seriously.
- Check her books and comics and point out the sort of female stereotypes which are passive and generally 'wet', and be equally critical of TV and magazine advertisements.
- Protect her self-image about maths and self-confidence from

the remarks of others. For example, challenge Uncle Jack when he implies that he wouldn't expect Jane to be any good at maths or science or to be interested in computing or engineering.
- Show as much interest and enthusiasm in maths as you can muster. After all, if you've got this far in the book you haven't done badly. Confidence is infectious and your daughter will learn from you that maths can be fun.
- Have a go with your daughter at some of the games and puzzles in this book. Look out for programmes on numeracy, maths, computers, science (and careers in all of these) on TV which you and she could watch together.
- Encourage her to play with calculators and computers. Remember there are many computer activities other than 'Space Invaders' to interest her. For example, the computer offers an exciting medium for constructing a quiz, drawing a picture or even composing a tune.

22
With a little help from my calculator

An understandable worry that is felt by many parents and teachers is that children will come to rely too much on the calculator and it will 'rot their brains'. (There are some teachers who insist that pupils are denied access to a calculator until they have achieved a secure mastery of number tables.) While this is an understandable fear, my own view is that this is a bit like not giving your child a bicycle until she has learnt to ride! It ignores completely the possibility that a calculator could actually *help* your child to learn her tables, and provide a lot more besides. For example, the calculator can:

- enable your child to tackle a wide range of everyday and mathematical problems that she could never otherwise attempt
- encourage her to explore mathematical patterns and functions just for the sheer fun of it
- let her concentrate her attention on particular mathematical ideas and relationships (for example, area or discovering the value of π) without getting bogged down in the arithmetic

You can rest assured that your child will gain more than she might lose if she has access to a calculator from an early age. Here is what the Cockcroft Report (reference 7, page 277) has to say:

> From all the studies the weight of evidence is strong that the use of calculators has not produced any adverse affect on basic computational ability.
>
> (Para. 377)

Much of my own professional work over the past few years has been concerned with the use of calculators in schools. This has involved the writing of calculator activities and games for children, working together with children and teachers to prepare in-service courses for teachers in the use of calculators in the classroom. I have been both excited and delighted by the results. The calculator has brought a new freshness into the teaching and learning of maths. Children who seemed to have lost their curiosity are now more prepared to explore maths through their calculator, knowing that it will never laugh at their silly mistakes or be cross if they get the wrong answer. They can tackle concepts knowing that they won't get stuck with the arithmetic. They can tackle real problems with realistic numbers. Most of all the calculator helps to stimulate children to pose their own questions about maths and builds their confidence to answer such questions.

Using a calculator sensibly

There was a time when computation (being able to do long calculations) was a valued and genuinely useful skill in our society. Clerks and cashiers spent hours performing 'long tots' and long-division calculations. The mathematics curriculum in schools reflected this by its emphasis on arithmetic. However, times have changed. Being good at computation no longer guarantees a job and it would be foolish to insist that pupils spend valuable time in school developing speed and accuracy in it. With the cheap availability of electronic calculators and computers we need to decide on certain priorities. Given that the hard slog of calculation will now be handled by a machine, what are the really useful *human* skills that we should be concentrating on? Here are a few which I feel have been made both more important and more accessible because of the calculator. These are also some of the essential skills needed if children are to use their calculators sensibly.

1. **Information** – knowing what information would help answer a problem.

2. **Units** – understanding the relevant units of measurement (is the price in pounds or pence?).
3. **Estimation** – knowing roughly what answer to expect.
4. **Calculation** – knowing what calculation to do (i.e., what key sequence to press on the calculator).
5. **Accuracy** – knowing how many figures of accuracy are relevant.

Clearly these skills have always been valued, but with a little help from the calculator we now have the opportunity to give children more practice in them than ever before.

This chapter should help to give you a basic grasp of what a calculator does as well as offer some suggestions for calculator activities and games which you can try with your child. I haven't recommended particular calculator models as there are so many on the market and they change almost daily. However, the popular models in schools tend to be Texas and Casio. Both these manufacturers offer a wide range of well-designed machines.

Now, when you have managed to find the ON switch on your own calculator, read on. . . .

Exploring your calculator

The keys on most calculators can be grouped into four main headings. These are known as the numbers, the operation keys, the function keys and finally the memory keys. There is also, of course, the $=$ key which you usually press at the end of a calculation to get the answer. Let's look at each of these four sorts of key starting with the numbers.

(i) *Number keys* These are simply the ten digits (0, 1, 2, . . . 9) and the decimal point, $\boxed{.}$. Most simple calculators will display numbers up to eight digits (including the decimal point). Here are a few questions you might like to explore on your calculator with your child.

WITH A LITTLE HELP FROM MY CALCULATOR   253

Exercise A

(i) What is the largest number you can enter into your calculator display?

(ii) What is the smallest *positive* number you can enter? (i.e., it must be greater than zero)

(iii) What appears on the display when you press the following?
 1 2 3 4 5 6 7 8 9

(iv) How do you enter a negative number like −25 into the calculator display?

(v) How do you clear a number from the display without having to switch the calculator off and on again?

COMMENTS ON THIS EXERCISE ARE GIVEN ON PAGE 262.

When your child reaches secondary school she may wish to buy a more sophisticated (and more expensive!) 'scientific' calculator. This will perform a few more tricks than the one you may have at present. For example, it can handle numbers very much bigger than 99 999 999 and smaller than 0.0000001. This is achieved by using what is known as 'scientific notation' and calculators which can work in this way will have a key marked either Exp or EE . If you possess such a calculator I'm afraid you'll have to look elsewhere for an explanation of how to enter and interpret numbers in this form.

(ii) *Operation keys* The four common operation keys are the familiar 'four rules' of ⊞, ⊟, ⊠ and ⊡. They are always used together with *two* numbers and the ⊟ key. For example, 2 ⊞ 4 ⊟, 10 ⊡ 2 ⊟, and so on.

Exercise B   Operation keys

Write down under 'Guess' the answers you would expect your

calculator to give for the following key sequences. Then press the sequences and check each answer in turn.

Key sequence	Guess	Press
(i) 2 ⊞ 3 ⊟		
(ii) 6 ⊟ 4 ⊟		
(iii) 4 ⊠ 3 ⊟		
(iv) 12 ⊟ 4 ⊟		
(v) 4 ⊠ ⊟		
(vi) 2 ⊟ 5 ⊟		
(vii) 2 ⊟ 3 ⊟		
(viii) 3 ⊞ ⊠ 2 ⊟		
(ix) 6 ⊟ ⊟ 2 ⊟		
(x) 2 ⊞ 3 ⊠ 4 ⊟		

COMMENTS ON THIS EXERCISE ARE GIVEN ON PAGE 262.

(iii) *Function keys.* Some basic calculators have no function keys at all. Others may have the square key ($\boxed{x^2}$) and perhaps the square root key (marked $\boxed{\sqrt{}}$ or $\boxed{\sqrt{x}}$). The function keys are used differently to the operation keys. If you play around with them you will discover that they don't need the $\boxed{=}$ key in order to give an answer. Also they act on only *one* number at a time; that is, the number which happens to be in display when you press the function key. Exercise C helps you to discover what these two function keys, $\boxed{x^2}$ and $\boxed{\sqrt{}}$, actually do.

Exercise C   *Function keys*

(a) Press the following sequences and write down the answer which they produce on your calculator display.

Key sequence	Answer
2 $\boxed{x^2}$	
3 $\boxed{x^2}$	
4 $\boxed{x^2}$	
4 $\boxed{\sqrt{}}$	
9 $\boxed{\sqrt{}}$	
16 $\boxed{\sqrt{}}$	
10 $\boxed{x^2}$ $\boxed{\sqrt{}}$	
10 $\boxed{\sqrt{}}$ $\boxed{x^2}$	

(b) Now write down what you think the $\boxed{x^2}$ and $\boxed{\sqrt{}}$ keys do to the number in display.

_____

_____

COMMENTS ON THIS EXERCISE ARE GIVEN ON PAGE 263.

If your calculator does not have a square key ($\boxed{x^2}$) you will almost certainly be able to square any number by the following method. You simply enter the number and then press $\boxed{\times}\boxed{=}$. Try it now and see if it works.

Try pressing sequences like:

2 $\boxed{\times}$ $\boxed{=}$

3 $\boxed{\times}$ $\boxed{=}$

and so on.

The more complicated scientific calculators offer a wide range of functions; for example, the trigonometric functions of sine, cosine and tangent, the logarithm function, and so on. These keys are just like the $\boxed{x^2}$ and $\boxed{\sqrt{}}$ keys in that they don't require the $\boxed{=}$ key and are pressed in the following sequence:

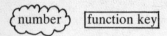

(iv) *Memory keys* Although many basic calculators have memory keys, they are rarely used because most people don't understand them. Have you ever been in the middle of a calculation and wanted to jot down one of the numbers you've worked out before going on to complete the calculation? Well, next time you want to do this, 'jot' the number down using your calculator's memory. The memory is simply a number storage system which allows you to store any number you choose. Later you can recall it to the display when you want to remind yourself what it is or use it in a calculation. This can be particularly useful in a calculation involving a particular number which is used repeatedly. Instead of entering the number each time you want to use it, simply store it in the calculator's memory and, when you need it, press 'memory recall'.

Unfortunately the labels for the memory keys vary from one calculator to another. Have a look at your calculator's manual and see how the memory keys work on your machine. Now try storing a number in the memory, clearing the display and then recalling the original number by pressing 'memory recall'. Once you are able to do this you have mastered the most important aspect of using the memory keys.

Finally, your calculator will have one or perhaps two keys which allow you to cancel all or part of a sequence that you have entered. These are usually marked $\boxed{C}$ or $\boxed{AC}$. The $\boxed{C}$ key is particularly useful if you wish to alter an incorrectly entered *number* in the middle of a calculation. Try this sequence:

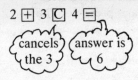

If you enter an incorrect *operation* key, you cancel it by pressing the correct operation key next in sequence. Try this sequence:

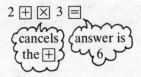

Learning maths with a calculator

One extremely useful feature which most calculators possess is what is known as their 'constant facility'. What this means is that calculators can be set up to perform the same calculation over and over again simply by pressing the ☐ key. Not only is this handy in certain types of calculation but, as you will see shortly, it can be a very powerful aid to learning maths and one that you can use with your child to good effect. Unfortunately the constant facility works for different calculators in different ways and indeed a few models do not possess it at all.

Now let's see if you can get your calculator's constant to work. Switch on and follow through the flow chart on page 222 with your calculator. If your machine *has* a constant facility you should finish up in one of the three clouds on the right hand side of the chart. Here is a brief explanation of the three main types of constant.

(a) ☒*key constant* If you want to multiply repeatedly by, say, 1.15, first set up the constant by pressing

   1.15 ☒ ☒

Then enter each number in turn and press $=$. Thus:

(Number 1) $=$

(Number 2) $=$

etc.

You will see that each number in turn is multiplied by 1.15.

NB *Don't press any other key* – the operation keys and the clear keys will override the constant that you have set up.

The same procedure is followed for constant calculations involving $+$, $-$ or $\div$.

(b) *Automatic constant* This is perhaps the most common constant found in basic calculators. The way it works is that once you have done one calculation, say

3.28 $\times$ 1.15 $=$

your calculator will continue to multiply by 1.15 any number you enter thereafter. All you have to do is to enter each further number and press $=$.

Again, *don't press any other key* than a number and $=$ or the constant will be lost. You may have to experiment with your own calculator to discover whether it uses as its constant the number *before* the operation or the one after it. This is why with some calculators the sequence:

3 $+$ 2 $=$ $=$

adopts the 2 as the constant and gives the answer 7, while others will adopt the 3 as the constant number and give the answer of 8.

(c) *Double press constant* This method is similar to the K key method. Instead of pressing 1.15 $\times$ K, you press 1.15 $\times$ $\times$. When the second $\times$ is pressed, a tiny 'K' may appear in the

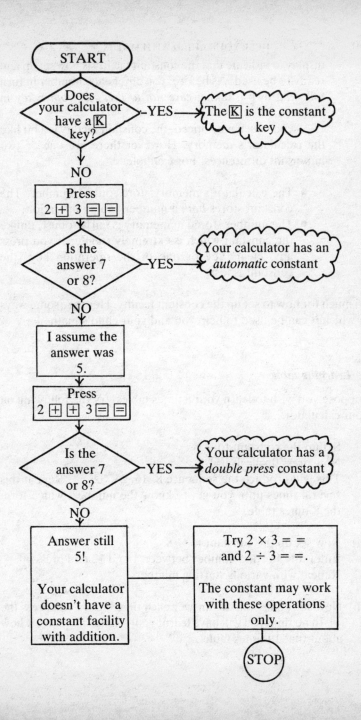

display to indicate that the constant facility is now set up and ready to be used. As before, you enter each number in turn followed by $=$, taking care *not to press any other key* in between.

As you may have noticed, the constant facility is a bit like the calculator's memory. However there are one or two important differences. For example:

- The calculator's memory stores only a number. The constant stores *both* a number and an operation.
- The number stored in memory is fairly robust, unlike the constant which is extremely fragile. If you press any 'clear' key or one of the operation keys the constant is lost.

So much for how to set up the constant facility. Here are some ways in which it can be used to help you and your child's maths.

1. *Learning tables*

Suppose you wish to learn your 8 times tables, try the following on your calculator:

(i) Set up the constant to + 8.
Now press $0 = = = =$
This will produce the sequence 8, 16, 24, 32 . . . Repeat this several times until you get to know the numbers which form the 8 times table.

(ii) Now set up the constant to × 8.
Enter any starting number between 1 and 12 and press $=$.
Repeat with various starting numbers.

(iii) Now start guessing the answer each time before you press the $=$. In no time you will have learnt your 8 times table. Now how about that 19 times table . . . ?

2. Discovering negative numbers

Set up your calculator's constant to −1.
Now enter the number 10 and with each press of ⌐=⌐ you can count down the numbers in display

   9, 8, 7, 6 . . .

But the calculator won't stop at zero (blast off). Keep pressing ⌐=⌐ and the display will continue into the world of negative numbers.

   . . . 0, −1, −2, −3, −4 . . .

3. Getting the hang of decimals

Set up your calculator's constant to + 0.1.
Now enter the starting number 0 and with each press of ⌐=⌐ you will get the sequence

   0.1, 0.2, 0.3, . . . 0.9, 1.0, 1.1 . . .

This can be very revealing for your child who may have expected that the number after 0.9 (nought point nine) would be 0.10 (nought point ten)!

Here are some other decimal constants to try:

   × 0.1 and ÷ 10 (they are the same)

   × 10 and ÷ 0.1 (they are the same)

4. Compound interest

You may have been taught a rather complicated formula for calculating compound interest at school. Both children and adults find this topic difficult, largely I think because they need to see how the interest grows year by year and this is difficult to calculate using pencil and paper. With a calculator, life is much simpler, but things are speeded up even more by using the calculator's constant facility. Try setting up your calculator's constant to × 2 and then start pressing ＝. You may be surprised how quickly the effect of doubling causes the calculator to reach a number bigger than it can handle.

Now how about a 10% increase each time you press ＝? To make a number 10% bigger you must multiply it by 1.10 (for 15%, use 1.15, and so on). Try setting up your calculator's constant to × 1.10 and see how it grows as you press ＝ repeatedly. This sort of playing with the constant facility can give your child a really valuable insight into how growth actually develops stage by stage. This in turn leads to better understanding of what happens to money in a bank and population growth.

ANSWERS TO EXERCISES FOR CHAPTER 22

Exercise A

Answers to the exercise will depend on whether you have a 'basic' or a 'scientific' calculator. The responses given below show the typical results from a 'basic' calculator.

(i)   999999999

(ii) 0.0000000

(iii) 12345678

(iv) If you have a ± key, press 25 ± . Otherwise, try 0 − 25 ＝ .

(v) Press C or CE .

Exercise B

Key sequence	Answer	Comment
(i) 2 + 3 =	5	The four rules in action.

(ii)	6 ⊟ 4 ⊟	2	
(iii)	4 ⊠ 3 ⊟	12	
(iv)	12 ⊡ 4 ⊟	3	
(v)	4 ⊠ ⊟	16	The number is squared.
(vi)	2 ⊟ 5 ⊟	−3	A negative number.
(vii)	2 ⊡ 3 ⊟	0.6666666	An unending decimal fraction.
(viii)	3 ⊞ ⊠ 2 ⊟	6	The first operation keys are ignored.
(ix)	6 ⊡ ⊟ 2 ⊟	4	
(x)	2 ⊞ 3 ⊠ 4 ⊟	20	One some calculators the ⊠ is done first giving the answer 14.

Exercise C

(a) 4; 9; 16; 2; 3; 4; 10; 10.

(b) x^2 squares the number on display.

$\sqrt{}$ is the square root key and is the opposite to x^2.

Calculator games

To end this chapter I have included four calculator games which have been used in many schools and have proved extremely successful in motivating children and giving them a better grasp of the maths skills on which they are based. Try some or all of them with your child. You will find that these games provide an excellent discussion point for some of the underlying maths skills. When the game gets too easy for her, ask your child to discuss the rules and adjust them to make the game more challenging. Perhaps she could introduce a handicapping system to make it harder for a better player. Finally you and your child may prefer to change the rules so that they are played collaboratively rather than competitively – for example, with Calculator Snooker you may try to make the biggest score possible working together. Have fun!

Calculator Snooker

For 2 players, aged eight-plus.

Player A enters any two-digit number. B takes a 'shot' by performing a multiplication sum. To 'pot' a ball, the first digit of the answer must be correct according to the table shown. (The degree of accuracy can be varied according to experience.)

Ball	Red	Yellow	Green	Brown	Blue	Pink	Black
Result needed	1...	2...	3...	4...	5...	6...	7...
Score	1	2	3	4	5	6	7

Otherwise, the rules are similar to 'real' snooker. There are 10 (or 15) reds and one of each of the six 'colours'. A player must score in the order red, colour, red, colour, and so on, until all the reds have gone. (Note that the colours are replaced but the reds are not.) When the last red has gone, the colours are potted 'in order' and are not replaced.

For example, one sequence of plays was:

Player	Enters	Display	Comments
Karen	69	69	
Peter	× 2 =	138	Peter pots the first red. He elects to go for blue ...
	× 5 =	690	... and misses.

Player	Enters	Display	Comments
Karen	× 2 =	1380	Karen pots the second red. She elects to go for black . . .
	× 5.5 =	7590	. . . and pots it.
	× 1.6 =	12144	The third red. She elects to go for black again . . .
	× 7 =	85008	. . . and misses.

Guess the Number

For 2 players (designed for calculators with a [K] key).

Player A chooses a number, say 40, and presses 40[÷][K] 0 (Note: the final 0 is pressed in order to clear the 40 from the display).

Player B has to guess which number A has chosen to divide by, by trying different numerators and pressing [=]. The aim is to guess A's number as quickly as possible.

Sample play: B's attempts to discover the denominator of 40 are as follows.

B presses	Display	Comments
1. 5 =	0.125	5 is too small
2. 24 =	0.6	24 is too small
3. 36 =	0.9	36 is too small
4. 40 =	1.	40 is the denominator

Arithmetrick

A game for 2–3 players, aged six years–adult.

Prepare a set of 21 cards numbered 0–20. Shuffle and turn them face down.

Player A enters any number (say 6) into the calculator. B draws the first card (say 10) and must perform the correct sum on the calculator ($\boxed{+}\,4\,\boxed{=}$) in order to win the card. It is then A's turn to draw a card (say 20). A must perform a sum starting with the 10 on the display of the calculator to get the value on the card. To encourage players to use $\boxed{\times}$ and $\boxed{\div}$ keys they get an extra turn if they use these successfully.

SAMPLE PLAY

Player	Display	Card	Sum	Tricks won A	B
B	6	10	$\boxed{+}\,4\,\boxed{=}$		1
A	10	20	$\boxed{\times}\,2\,\boxed{=}$	1	
A	20	11	$\boxed{-}\,9\,\boxed{=}$	1	
B	11	9	$\boxed{-}\,2\,\boxed{=}$		1
A	9	etc.			

10   20   11   9

Maths skills	Reinforces number bonds. Encourages the children to move from the 'safe' operations of $\boxed{+}$ and $\boxed{-}$ to the more complex $\boxed{\times}$ and $\boxed{\div}$.

Extending the game for older children

Even eight- and nine-year-olds will soon discover that they can go from, say, a 10 to a 15 by multiplying by 1½. It isn't long before they learn these common decimal fractions:

½ = 0.5; ¼ = 0.25; ¾ = 0.75

What about ⅓, ⅖, ⅜? Perhaps you'd better explore these yourself!

Of course you are not restricted to the numbers 0–20 or even to the four operations +, −, × and ÷. Look at the other keys you have on your calculator and see if you can include them.

Space Invaders

This game can be played at different levels (1, 2, 3, etc.). Move on to a new level when you find the game too easy.

Space Invaders 1

Enter a three-digit number into the calculator (say 352). These three digits are the aliens; you shoot them down one at a time by subtracting to zero.

Example: Starting number 352

Key presses	Display
⊟ 2 ⊟	350
⊟ 5 0 ⊟	300
⊟ 3 0 0 ⊟	0

You can make up your own numbers and let your partner shoot them down.

Try the following: 416, 143, 385, 512, 853, 264, 179, 954, 589, 741.

Space Invaders 2

The same as Space Invaders 1, except that the digits must be shot down in ascending order.

Example: Starting number 352

Key presses	Display	
⊟ 2 ⊟	350	i.e., you shoot down the 2, then the 3, and then the 5.
⊟ 3 0 0 ⊟	50	
⊟ 5 0 ⊟	0	

Space Invaders 3

The same as Space Invaders 2, except that you can use numbers with more digits. Try 4-digit numbers, then 5-, 6-, 7- and 8-digit ones.

Space Invaders 4

The same as Space Invaders 3, except that you shoot the digits down by addition, not subtraction. Use as few shots as possible.

Example: Starting number 1736

Key presses	Display	
⊞ 4 ⊟	1740	If you start with a 5-digit number, the game ends with a display 100000.
⊞ 6 0 ⊟	1800	
⊞ 2 0 0 ⊟	2000	
⊞ 8 0 0 0 ⊟	10000	

Space Invaders 5

The same as Space Invaders 3, except that you can use decimals, e.g., 451.326 to be shot down by subtraction in the order 1, 2, 3, 4, 5, 6.

Space Invaders 6

The same as Space Invaders 5 except with the added constraint that each digit can only be shot down in the digits column. Thus the

WITH A LITTLE HELP FROM MY CALCULATOR 269

display value must be multiplied or divided by 10,100,1000 . . . in order to get the appropriate digit into the unit position each time.

ANSWERS TO PRACTICE EXERCISES

Chapter 6, page 72

1. Weekly milk bill = 21 × 23 pence (number of pints per week)

    = 483 pence

    Change from £5 = 500 − 483 pence
    = *17 pence*

2. Number of stamps = $^{100}/_{17}$
    = 5 stamps and 15p change.

3. One fifth is the same as 20% and so is a better reduction than 5%.

4.
METER READINGS

Present	Previous	Units supplied	Pence per unit	Amount (£)
56769	56242	527	5.10p	26.88
			Fixed charges	£4.64
			Total	£31.52

Chapter 7, page 81

1. No comment.

2.

Number	365	614	496	16042	1093461
Place value	10	100	1	1000	10

3. No comment.

4.

Temperature °C	10	4	21	−6	−10	0	3	−3
Three degrees less	7	1	18	−9	−13	−3	0	−6

5. Calculations where you add (+) or multiply (×) can be done in *any* order. However, those where you subtract (−) or divide (÷) will give a different answer depending which way round you have the numbers.

Chapter 9, page 102

1. 2¾, 3⅖, ⅚, 6⅔

2.
```
   1 = 3 thirds        10 = 30 thirds
   4 = 12 thirds       3⅓ = 10 thirds
   5 = 15 thirds       7⅔ = 23 thirds
```

WITH A LITTLE HELP FROM MY CALCULATOR

3.

$$¼ + ⅙ + 1/12 + ½$$
$$↓ \quad ↓ \quad ↓ \quad ↓$$
$$3/12 + 2/12 + 1/12 + 6/12$$
$$↘ \quad ↓ \quad ↙$$
$$12/12 = 1$$

4.

Fractions	8/10	4/6	5/10	12/18	6/9	4/16	8/48	9/18	2/22
Simple form	4/5	2/3	1/2	2/3	2/3	1/4	1/6	1/2	

5.

Fractions		2/3	3/4	2/6	7/12	5/6	1/2
Fractions as twelfths		8/12	9/12	4/12	7/12	10/12	6/12
Rank		3	2	6	4	1	5

Chapter 10, page 114

1. Mark the numbers 0.35 and 0.4 on the number line below.

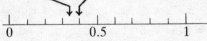

   Which of the two numbers is bigger? _____0.4_____

2. In 0.6, the 6 stands for 6 tenths

3. Ring the number nearest in size to 0.78

   0.7 / 70 / 0.8 / 80 / .08 / 7

4. Multiply by 10: 5.49 →   54.9

5. Add one tenth: 4.9 →   5.0

6.

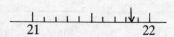

   This number is about   21.85

7. There is no limit to the number of numbers that exist between 0.26 and 0.27. Here are just a few:

   0.261, 0.262, 0.2614, 0.2683172 . . . etc.

Chapter 11, page 127

1. The fraction ⅛ = 0.125 or 12½%.
   Therefore ⅛ is *bigger* than 8%.

2. The fraction ¹⁄₁₅ = 0.0666 . . . or about 7%.
   Therefore ¹⁄₁₅ is *less* than 15%.

3. Even though the rate of inflation has fallen, prices are still rising by 4% per year. So the answer is (a).

4. 20% = ⅕
   ⅕ of £80 = £16

5. £1.60.

6. The actual price rise is the same for both sizes of eggs. However, the percentage price rise is greater for the cheaper (size 5) eggs.

7. (a) 20.93 ÷ 1.15 = £18.20
   (b) 18.20 × 1.25 = £22.75

8. The reason that children's sweets suffer greater inflation than other goods is because any price rise, however small, will represent a very large increase in percentage terms. For example, the price of a 5p toffee bar could not go up by less than 1p, which would mean a 20 per cent increase in its price.

Chapter 14, page 157

Exercise 1

(1) 4125 (2) 38.42 (3) 291.7 (4) 39040
(5) 39050 (6) 38.41 (7) 447.0 (8) 0.1429
(9) 1318 (10) 3050.

Exercise 2

Problem	Answer
1	4
2	3
3	3.33
4	3⅓

Chapter 15, page 174

1. (i)   60 kg: (about 9½ stones)

(ii) 56 lbs: 25 kg is only 55 lbs
(iii) Probably about 2.05 m
(iv) 90 kmph is about 60 mph
(v) ½ litre is less than one pint (actually 0.88 pints)
(vi) 4½ half-litre bottles is about 4 pints
(vii) 100 g is just under half of a ½ lb packet.

Chapter 16, page 190
1. No comment.

2. (i) 90   (ii) 4   (iii) 180
   (iv) 60   (v) square   (vi) 165*

* When the hand of a clock goes right round the clock face it turns through an angle of 360°. Thus in one hour, when it moves between 12 and 1, the hour hand turns through 30° ($^{360}/_{12}$). Between twelve o'clock and half past twelve the small hand turns through an angle of 15° while the big hand turns through 180°. Since the two hands started together at twelve o'clock the angle between them must be 165° at twelve thirty.

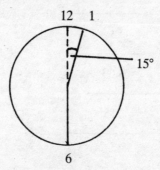

3. Wheel circumference = 70 π cm
   Number of turns   = 100,000/70π
                      = 454.7

4. 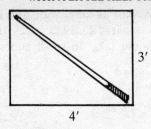 Make a parcel 3′ by 4′. The cue should fit exactly into the diagonal!

Chapter 17, page 203

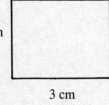
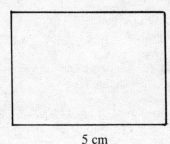

2. The triangles ADE and ABC have the same shape and so their sides are in the ratio 2:1. Since DE is twice BC, it must be 8 units long.

3.

Ingredients	Amounts for 15 servings
Toasted hazelnuts	560 g
Cornflour	5 tablespoons
Separated eggs	5 eggs
Castor sugar	190 g
Milk	750 ml
Vanilla essence	about 8–10 drops
Double cream	750 ml

4. If commercial sugar cubes are half as long, wide and high as domestic ones, their volume is ⅛ that of domestic cubes (this is ½ × ½ × ½). Therefore, I should take 16 small cubes to be equivalent to 2 large ones!

References

1. Glenn Doman, *Teach Your Baby Maths* (Jonathan Cape), London, 1979.
2. Boris A. Kordemsky, *The Moscow Puzzles* (Pelican), Bungay, Suffolk, 1978.
3. Martin Gardner, *Mathematical Puzzles and Diversions* (Pelican), London, 1976.
4. K. M. Hart, ed., *Children's Understanding of Mathematics: 11–16* (Murray), Oxford, 1981.
5. Linda M. Walker, *Origami: Paper-folding Made Easy* (John Bartholomew and Son Ltd), Edinburgh, 1974.
6. *Make Shapes (Series No.1)* (Tarquin Publications), Diss, Norfolk.
7. *Mathematics Counts* – Report of the Committee of Inquiry into the Teaching of Mathematics in Schools under the Chairmanship of Dr W. H. Cockcroft (HMSO), London, 1982.

Index

Index

add 75
algebra 67–9, 227–38
and 75
angle 151, 180–3
area 151, 168–9, 188
arithmetic 67–9
Arithmetrick 266–7

bar charts 223

cake diagram 97–100
calculator games 263–9
Calculator Snooker 264–5
calculator's constant facility 118, 257–62
capacity 151, 172
cardinal numbers 43–5, 73, 83
circle 34, 185–8
circumference 185–6
classifying 36–7, 221
colour 32
comparing 23, 54–6
composite numbers 80
compound interest 261–2
cone 189
conservation of number 41
counting 23, 39–52
counting skills 46–8
cube 189
cuboid 189
cylinder 189

decimal point 112
decimals 108–19
denominator 101
describing 22–4
diagrams 220–2
diameter 185–6
difference 75
digits 77
dimensions of measure 151
divide 75

ellipse 34
equilateral triangles 180, 215
estimate 156
even numbers 80

factors 80
Feelies 35
Fibonacci sequence 207–8
Finger tables 90
flow chart 222
four rules 75–6, 112–15
fractions 95–107
    common 95
    decimal 95, 108
    equivalent 99, 121
function keys (on a calculator) 254–6

geometry 67–9
goes into 75
Golden ratio 209–12
graphs 222–6
Guess my Number 90
Guess the Number 265

hexagons 34, 180, 216
hypotenuse 185

imperial units 164
In the Bag 27
isosceles triangle 215

length 151, 165–8
line graphs 224
logic 70
lowest common denominator 101

Magic squares 91
matching 28–31
measuring 58–60, 150–63
memory keys (on a calculator) 256–7
metric units 164
minus 75, 78
multiply 75
multiplying fractions 102

naming 22, 26–8
net 189
network 222
number 55
  negative numbers 78
  whole numbers 73–88

number keys (on a calculator) 252–3
number line 74, 98, 110, 122, 129
number plate games 89
number tables 84
numerals 49
nurture 241

odd numbers 80
Odd One Out 35
one-to-one correspondence 46, 47–8
operation keys (on a calculator) 253–4
operations 75
order 23, 56–8, 222
ordering 56–8
ordering scale 154
ordinal numbers 43–5, 74, 83

patterns 207–19
pentagon 34, 179, 180, 216
percentages 120–31
pi 187–8
pie charts 223
place value 77, 117
plus 75
position 55
positive numbers 78
posting toys 31
prime numbers 80
problem-solving stages 142–3
product 75
proof 234–5
proportion 195–206
Pub Cricket 89
Pythagoras' theorem 185

quadrilateral 180, 215, 216

radius 185–6
ratio 197, 210
rectangle 34
rectangular numbers 80
representing numbers 48–50
right angle 181
rounding 146, 156, 157

scale factor 201
scalene triangle 215
scatter graphs 224
sector (of a circle) 34
sequence 56–7
set 22, 73
shape 32–5, 179–88
share 75
significant figures 155–6, 163
size 32, 55
solids 188–93
sorting 22, 32–6
space 70, 179
Space Invaders 77, 267–9
speed 151
sphere 189
square 34
square numbers 80
sum 75

take away 75
temperature 151
tessellation 213
time 151
times 75
times tables 84
trapezium 34

tree diagrams 221
triangle 34, 180, 182–5

VAT calculations 124, 126
Venn diagrams 220–1
volume 151, 169–72

weight 151

Fontana Paperbacks: Non-fiction

Fontana is a leading paperback publisher of non-fiction, both popular and academic. Below are some recent titles.

- ☐ CAPITALISM SINCE WORLD WAR II Philip Armstrong, Andrew Glyn and John Harrison £4.95
- ☐ ARISTOCRATS Robert Lacey £3.95
- ☐ PECULIAR PEOPLE Patrick Donovan £1.75
- ☐ A JOURNEY IN LADAKH Andrew Harvey £2.50
- ☐ ON THE PERIMETER Caroline Blackwood £1.95
- ☐ YOUNG CHILDREN LEARNING Barbara Tizard and Martin Hughes £2.95
- ☐ THE TRANQUILLIZER TRAP Joy Melville £1.95
- ☐ LIVING IN OVERDRIVE Clive Wood £2.50
- ☐ MIND AND MEDIA Patricia Marks Greenfield £2.50
- ☐ BETTER PROGRAMMING FOR YOUR COMMODORE 64 Henry Mullish and Dov Kruger £3.95
- ☐ NEW ADVENTURE SYSTEMS FOR THE SPECTRUM S. Robert Speel £3.95
- ☐ POLICEMAN'S PRELUDE Harry Cole £1.50
- ☐ SAS: THE JUNGLE FRONTIER Peter Dickens £2.50
- ☐ HOW TO WATCH CRICKET John Arlott £1.95
- ☐ SBS: THE INVISIBLE RAIDERS James Ladd £1.95
- ☐ THE NEW SOCIOLOGY OF MODERN BRITAIN Eric Butterworth and David Weir (eds.) £2.50
- ☐ BENNY John Burrowes £1.95
- ☐ ADORNO Martin Jay £2.50
- ☐ STRATEGY AND DIPLOMACY Paul Kennedy £3.95
- ☐ BEDSIDE SNOOKER Ray Reardon £2.95

You can buy Fontana paperbacks at your local bookshop or newsagent. Or you can order them from Fontana Paperbacks, Cash Sales Department, Box 29, Douglas, Isle of Man. Please send a cheque, postal or money order (not currency) worth the purchase price plus 15p per book for postage (maximum postage required is £3).

NAME (Block letters) _____

ADDRESS _____ _____

While every effort is made to keep prices low, it is sometimes necessary to increase prices at short notice. Fontana Paperbacks reserve the right to show new retail prices on covers which may differ from those previously advertised in the text or elsewhere.